Lead-free Piezo-Ceramic Solid Solutions

by

Dr. R. Saravanan, M.Sc., M.Phil., Ph.D.

Associate Professor & Head

Research Centre and PG Department of Physics

The Madura College (Autonomous)

Madurai - 625 011

India

Published by **Materials Research Forum LLC**
Millersville, PA 17551, USA

Published as part of the book series
Materials Research Foundations
Volume 41 (2018)
ISSN 2471-8890 (Print)
ISSN 2471-8904 (Online)

Print ISBN 978-1-945291-94-4
ePDF ISBN 978-1-945291-95-1

Distributed worldwide by

Materials Research Forum LLC
105 Springdale Lane
Millersville, PA 17551
USA
http://www.mrforum.com

Manufactured in the United States of America
10 9 8 7 6 5 4 3 2 1

Table of Contents

Preface

Preface

Piezoelectric materials with perovskite and related structures have been studied extensively on fundamental and technological aspects. Piezoelectric materials are commonly used in sensor and actuator technologies due to their unique ability to couple electrical and mechanical displacements, *i.e.*, to change electrical polarization in response to an applied mechanical stress or mechanical strain in response to an applied electric field. The lead-based perovskite ceramics are dominating the markets of both the piezoelectric and the ferroelectric devices, but the toxicity of lead has raised serious environmental issues. Hence, the growing demand for replacing these lead-based piezoelectrics has given the need to focus on the development of efficient lead-free systems.

During the last several years, new lead-free piezoelectric materials have been developed to replace the lead-based materials. Presently, the families of lead-free ceramics showing most promising piezoelectric properties are based on $BaTiO_3$ modified sodium potassium niobate, sodium bismuth titanate, *etc*. In this context, a systematic investigation has been undertaken to analyze the dielectric, ferroelectric and piezoelectric properties of lead-free solid solution ceramics. All the compounds reported in this book have been prepared by solid-state reaction method. Several experimental techniques have been employed to characterize the solid solution ceramics. Charge density picture has been analyzed qualitatively and quantitatively for the type of bonding existing in solids. The present book deals with the electron density distribution studies of five series of $BaTiO_3$ based lead-free piezoelectric ceramics using experimental X-ray diffraction data. An investigation of the results obtained from other characterization works like UV-visible spectroscopy, scanning electron microscopy, energy dispersive X-ray spectra, dielectric spectroscopy, dielectric, ferroelectric, and piezoelectric study has also been carried out in this book. This book consists of five chapters.

Chapter I: Introduction

This chapter gives the objectives of this book, information about piezoelectric and ferroelectric materials, structural properties and applications of the lead-free piezoceramics. This chapter explains the basic information about PXRD profile fitting technique in a detailed manner. Also, this chapter discusses the methodology for calculating the experimental charge density. The most reliable model to estimate the charge density distribution is maximum entropy method which is explained in this chapter in a detailed manner with corresponding methodology.

Chapter II: Synthesis and Experimental techniques

This chapter presents sample preparation methods for five piezoceramics. It provides and discusses the working principles of the characterization techniques such as, powder X-ray diffraction, scanning electron microscopy, energy dispersive X-ray spectroscopy, UV-visible spectroscopy, dielectric, ferroelectric and piezoelectric measurements.

Chapter III: Results

Chapter III presents the results obtained from various characterization techniques and analytical methods performed to investigate the five differently doped $BaTiO_3$ lead-free ceramics systems. The plots of experimental X-ray diffraction patterns, Rietveld fitted profiles, SEM micrographs, EDS spectra, Tauc plots, dielectric, ferroelectric measurements, two dimensional, three dimensional and one dimensional electron density line profiles are presented. The tables of structural parameters refined from Rietveld method and optical band gap values with respect to the dopant concentration are also given. The elemental compositions of the prepared ceramics from EDS analysis, parameters of dielectric, ferroelectric and piezoelectric measurements, bond lengths and mid-bond electron density values from MEM analysis are also presented.

Chapter IV: Analysis of results

Chapter IV provides the detailed analysis of results obtained for all the lead-free solid solutions. Interpretation of results, comparison of physical properties and electron density distribution for all the prepared solid solutions are also discussed.

Chapter V: Conclusions

This chapter gives the conclusion of the findings of the work reported in this book.

The results of the analysis on the piezoelectric ceramics reported in this book have been previously published as follows:

1. Structure and charge density properties of $(1-x)(Na_{1-y}K_yNbO_3)-xBaTiO_3$ lead-free ceramic solid solution, *S. Sasikumar,* R. Saravanan, Journal of Electronic Materials, Springer, 46, 4187-4196, (2017)

2. Piezoelectric and ferroelectric properties of lead-free $(1-x)(Na_{1-y}K_y)$ $(Nb_{1-z}Sb_z)O_3-xBaTiO_3$ solid solution, *S. Sasikumar,* R. Saravanan, K. Aravinth, Physica B: Condensed Matter, Elsevier, 512, 58-67, (2017)

3. Charge correlation of ferroelectric and piezoelectric properties of $(1-x)(Na_{0.5}Bi_{0.5})TiO_3-xBaTiO_3$ lead-free ceramic solid solution, *S. Sasikumar,* R.

Saravanan, S. Saravanakumar, K. Aravinth, Journal Materials Science: Materials in Electronics, Springer, 28, 9950-9963, (2017)

4. Preparation, electronic structure and chemical bonding of lead-free $(1-x)(K_{0.5}Bi_{0.5})TiO_3-xBaTiO_3$ solid solution, *S. Sasikumar,* R. Saravanan, S. Saravanakumar, M. Charles Robert, Applied Physics A: Materials Processing, Springer, 124, 1-10, (2018)

5. Investigation on charge density, piezoelectric and ferroelectric properties of $(1-x)Ba(Zr_{0.2}Ti_{0.8})O_3-x(Ba_{0.7}Ca_{0.3})TiO_3$ lead-free piezoceramics, *S. Sasikumar,* R. Saravanan, S. Saravanakumar, Journal Materials Science: Materials in Electronics, Springer, 29, 1198-1208, (2018)

6. Piezoelectric and Ferroelectric Properties of Lead-free $0.9(Na_{0.97}K_{0.03}NbO_3)$ -$0.1BaTiO_3$ Solid Solution, *S. Sasikumar,* R. Saravanan, S. Saravanakumar, Mechanics, Materials Science & Engineering, (Magnolithe GmbH), (2017), ISSN 2412-5954, DOI: 10.2412/mmse.47.30.332

Lead-free Piezo-Ceramic Solid Solutions, R. Saravanan   Materials Research Forum LLC
Materials Research Foundations **41** (2018)   doi: http://dx.doi.org/10.21741/9781945291951

Chapter 1

Introduction

Abstract

This chapter gives the objectives of this book, information about piezoelectric and ferroelectric materials, structural properties and applications of the lead-free piezoceramics. This chapter explains the basic information about the PXRD profile fitting technique in a detailed manner. Also, this chapter discusses the methodology for calculating the experimental charge density. The most reliable model to estimate the charge density distribution is the maximum entropy method which is explained in this chapter in a detailed manner with corresponding methodology.

Keywords: Piezoelectric, Dielectric, Ferro Electric, Fourier, MEM, Rietveld, Lead-Free, Sodium-Potassium Niobate, Sodium Bismuth Niobate, Potassium Bismuth Niobate, Barium Titanate, Solid Solution

Contents

1.1   Objectives

The objective of the work reported in this book is to synthesize and characterize some lead-free piezo-ceramics. The lead-based ceramic materials are well known for their excellent piezoelectric properties, which are widely used in industrial applications. The toxicity of lead-based piezoceramics has led researchers to develop new, environmental friendly piezoelectric materials as potential replacement for lead-based materials. In this context, the following five series of lead-free piezoceramic (solid solutions) have been chosen for the work reported in this book.

(i) $(1-x)(Na_{1-y}K_y)NbO_3$-$xBaTiO_3$ (x=0.1, 0.2; y=0.01, 0.05) (NKN-BT)

(ii) $(1-x)(Na_{1-y}K_y)(Nb_{1-z}Sb_z)O_3$-$xBaTiO_3$ (x=0.1, 0.2; y=0.03, 0.05; z=0.05, 0.1) (NKNS-BT)

(iii) $(1-x)(Na_{0.5}Bi_{0.5})TiO_3$-$xBaTiO_3$ (x=0.00, 0.04, 0.08, 0.12) (NBT-BT)

(iv) $(1-x)(K_{0.5}Bi_{0.5})TiO_3-xBaTiO_3$ (x=0.00, 0.08, 0.12) (KBT-BT)

(v) $(1-x)Ba(Zr_{0.2}Ti_{0.8})O_3-x(Ba_{0.7}Ca_{0.3})TiO_3$ (x=0.4, 0.5, 0.6) (BZT-BCT)

A brief description of the objectives of the present book is given as follows;

(i)   To synthesize the lead-free solid solutions by solid-state reaction method (SSR).

(ii)   To investigate the synthesized lead-free piezoceramics by powder X-ray diffraction for structural analysis.

(iii)   To analyze the surface morphological and microstructure properties of the lead-free piezoceramics by scanning electron microscopy (SEM).

(iv)   To analyze the elemental compositions of the materials qualitatively and quantitatively using energy dispersive X-ray spectroscopy (EDS).

(v)   To determine the optical band gap (E_g) for all the materials by UV-visible absorption spectra employing Tauc plot technique (Wood and Tauc, 1972).

(vi)   To study the variation of dielectric constant (ε) and dielectric loss tangent (tan δ) at different frequencies (1 kHz to 1MHz) as a function of temperature for all the solid solution systems using dielectric measurements.

(vii)   To analyze the ferroelectric properties of the prepared solid solutions using P-E polarization versus electric filed) hysteresis loop measurements.

(viii)   To analyze the piezoelectric properties of the prepared samples by measuring the piezoelectric constant (d_{33}) using a piezo-meter.

(ix)   To analyze the charge density distributions and bonding features using maximum entropy method (MEM) (Collins, 1982), employing experimental X-ray data.

(x)   To correlate the charge derived properties of the lead-free piezoelectric ceramics with the addition of the dopant with ferroelectric and piezoelectric properties.

The following sections provide a brief description about dielectric materials and the phenomenon of ferroelectricity and piezoelectricity.

Materials Research Forum LLC
doi: http://dx.doi.org/10.21741/9781945291951

1.2 Fundamentals of dielectric materials

1.2.1 Dielectrics

Ceramic materials that are good electrical insulators are referred to as dielectric materials. A dielectric is an electrically insulating material in which the atoms/ions can be polarized by the application of an external electric field. The polarized structure of dielectric material is the result of separation of positive and negative charge centers. These charge displacements are known as dipole moments and figure 1.1 illustrates this phenomenon. The elementary dipoles in the dielectric materials interact with each other under certain thermodynamic conditions. Due to this electric polarization phenomena, dielectric materials acquire opposite charges across their surfaces and are utilized for charge storage in capacitor applications. The electric dipole moment per unit volume of a dielectric material is known as electric polarization and is proportional to the external applied electric field for linear dielectrics. The net macroscopic polarization (P) is related to relative permittivity/dielectric constant (ε_r) through the equation,

$$P = \varepsilon_0 E \, (\varepsilon_r - 1) = \varepsilon_0 \chi \, E \tag{1.1}$$

where, ε_0 is the permittivity of free space (8.854×10^{-12} F/m), E is the applied electric filed and $\chi = (\varepsilon_r - 1)$ is the dielectric susceptibility. The macroscopic polarization can also be expressed as a summation of all polarizations originating at the microscopic level.

In dielectrics, there are mainly four types of polarization mechanisms contributing to relative permittivity. The various mechanisms of electric polarization in a dielectric medium (low conducting at moderate electric fields) can be categorized as, atomic, ionic, orientation and space charge polarization (Moulson and Herbert , 2003).

(i)   Atomic polarization is based on the relative shift of negative charges (electron cloud) and positively charged nucleus of the atom in the external electric field.

(ii)   Ionic polarization takes place by the relative displacement of positive and negative ions in the presence of an applied electric field.

(iii)   Orientational polarization deals with the alignment of permanent dipoles.

(iv)   Space charge polarization deals the charge carriers displaced by an electric field and is stopped at the surfaces or grain boundaries of the material. This movement of charge carriers to the surface creates dipoles.

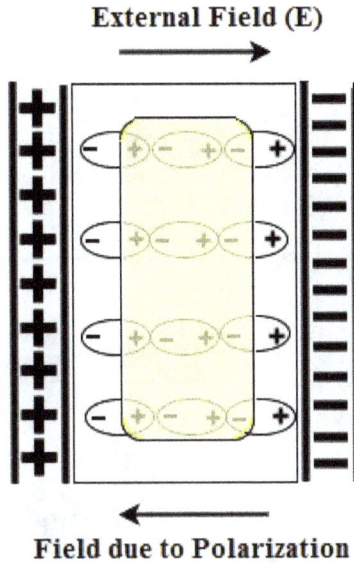

Figure 1.1 *Polarization of a dielectric material under static electric field.*

1.2.2 Classification of dielectrics

All crystalline materials can be represented by 32 crystallographic point groups as shown in figure 1.2. These point groups are divided into two classes which include a centre of symmetry (11 point groups) and non-centric symmetry (21 point groups). There are 21 non-centrosymmetric point groups, 20 (except point group 432) of which exhibit piezoelectricity. The other 11 centrosymmetric point groups are non-piezoelectric materials. Piezoelectrics are again divided into pyroelectrics (10 classes) possessing spontaneous polarization (P_s) and non-pyroelectrics (10 classes) that do not possess spontaneous polarization (Gene, 1999). Figure 1.2 shows that there are 10 crystal classes, out of a possible 20 that are designated as pyroelectric. This group of materials possess permanent dipoles within a given temperature range. A subgroup of the spontaneously polarized pyroelectrics is a very special category of materials known as ferroelectrics. Similar to pyroelectrics, the ferroelectric materials possess spontaneous polarization.

Materials Research Forum LLC
doi: http://dx.doi.org/10.21741/9781945291951

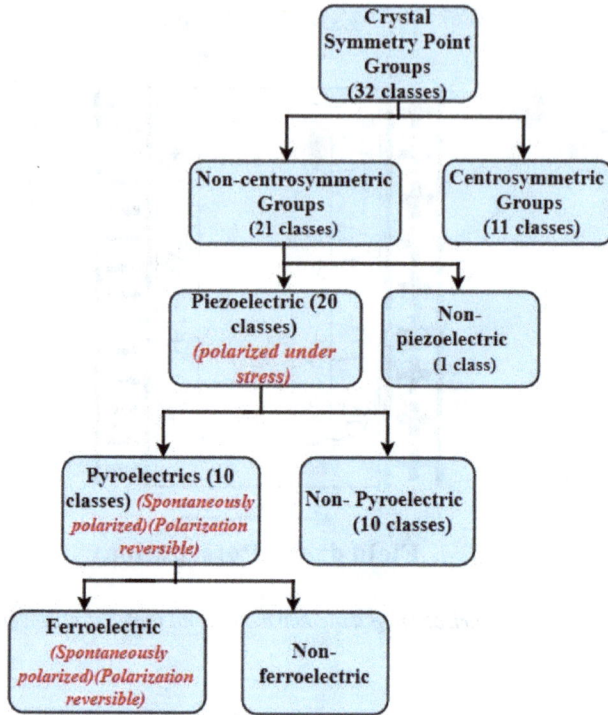

Figure 1.2 Classification of point groups on the basis of symmetry and piezoelectric, pyroelectric and ferroelectric properties.

1.2.3 Ferroelectricity

Ferroelectricity is characterized by the existence of spontaneous polarization in the absence of an electric field. The spontaneous polarization can be switched by application an external electric field (Spaldin, 2007). Ferroelectric materials undergo a structural phase transition from a paraelectric phase to a ferroelectric phase upon cooling through the Curie temperature (T_C). Below the Curie temperature (T_C), the crystal exhibits ferroelectricity and has a structure resulting from a change in the symmetry of the unit cell. Above T_C, the crystal changes to a centrosymmetric structure and the crystal has no spontaneous polarization. When the ferroelectric material is cooled below T_C, the central ion in the unit cell displaces from its equilibrium position and creates a spontaneous polarization. Figures 1.3 (a)-(c) show the orientations of the dipoles in a ferroelectric material with respect to external electric field.

1.2.3.1 Ferroelectric hysteresis

In ferroelectric materials, regions of aligned dipoles are referred as domains and the boundary separating these domains is known as domain wall. When ferroelectric materials are subjected to an electric field at a temperature below the Curie point, the domains start aligning in the direction of the field. For small magnitude of electric field, the effect is linear due to non-alignment of all domains. As the field strength is increased, a nonlinear increase in polarization occurs until it becomes saturated. At the saturated stage, the polarization is known as saturation polarization (P_s). The saturation polarization (P_s) at this stage is usually taken as an intercept on the polarization axis extrapolating from the tangent on saturation polarization (Damjanovic, 1998).

When the magnitude of the electric field is decreased gradually to zero, the polarization starts to decline but polarization does not tend to zero due to locking of some domains in the direction of the field, termed as remnant polarization (P_r). To get zero polarization, a field of opposite polarity with specific strength is required, called coercive field (E_c). Further increase of field strength in the negative direction initiates a new alignment of domains. Thus, the curve traced out as a result of the alignment and misalignment of domains in positive and negative directions of field is known as hysteresis curve or P-E curve (Figure 1.4).

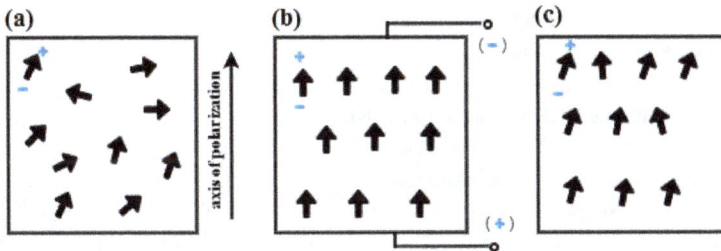

Figure 1.3 Orientation of (a) initial state without external electric field, (b) polarization in external electric field, (c) on removing the electric field, some crystallites revert to more energetically favourable positions.

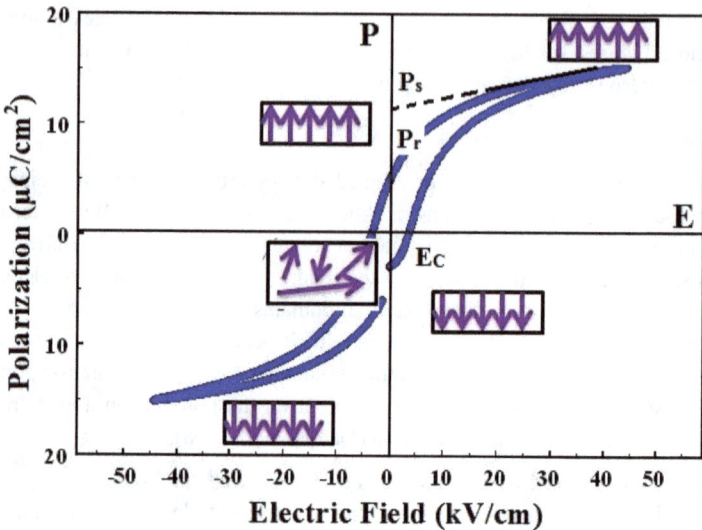

Figure 1.4 *P-E hysteresis loop for a polycrystalline ferroelectric material* (P_s - *saturation polarization,* P_r - *remnant polarization,* E_C - *coercive field*).

1.2.4 Piezoelectricity

Piezoelectricity (or direct piezoelectric effect) is the ability of some materials to generate electric charges or electric potential in response to applied mechanical stress (Segal, 1991). When a stress is applied on a piezoelectric material, one side of the material derives a net positive charge and the opposite side derives a net negative charge. Only certain materials that are anisotropic with non-centric symmetry are piezoelectric in nature (among the 32 crystal classes, 20 are piezoelectric). Conversely, a mechanical deformation produced in the same material with the application of electric field, is known as the converse piezoelectric effect (Damjanovic, 1998; Richerson, 1992). Materials exhibiting both these types of phenomena are considered piezoelectric materials and widely used for variety of applications. The two types of piezoelectric effects can be described by the following relations (Moulson and Herbert, 2003);

$$D = dT + \varepsilon^T E \quad \text{(Direct piezoelectric effect)} \tag{1.2}$$

$$S = S^E\, T + dE \quad \text{(Converse piezoelectric effect)} \tag{1.3}$$

Where D is the electric displacement, directly related to the polarization of the material, T is the stress, E is the electric field, S is the strain, d is the piezoelectric coefficient that has a unit of coulombs/Newton, and ε^T is the dielectric constant while the stress remains constant. S^E is the elastic compliance and remains constant under the electric field.

1.2.5 Perovskite structures

The most studied piezoelectric materials are perovskite type structures. A perovskite structure has a general formula ABO_3, where A, B are cations and O is an anion. The unit cell of the oxide perovskite $BaTiO_3$ is cubic structure and represented with the corners occupied by a large cation (A-site, such as Ba), a smaller cation in the center of unit cell (B-site, such as Ti) and oxygens in the face centers, as shown in figure 1.5. The lattice constant of perovskite structure is always close to 4 Å due to the rigidity of the oxygen octahedra network and the well defined oxygen ionic radius of 1.35 Å. Goldschmidt (Goldschmidt, 1926) introduced an empirical relation as given in equation (1.4) to describe ideal perovskite structure based on the packing of ions as rigid spheres. The perovskite structure can accommodate a variety of substituent ions with different ionic radii to the host structure. The stability of the structure depends on the ionic radii which is explained using tolerance factor (t) introduced by Goldschmidt, (1926) given by the following relationship:

$$t = \frac{R_A + R_O}{\sqrt{2}(R_B + R_O)} \tag{1.4}$$

where R_A, R_B and R_O are the respective ionic radii of A, B cations and oxygen. Based on the Shannon ionic radii (Shannon, 1976), for stable perovskite structures, a perovskite structure may exist for a tolerance factor in the range of $0.88 < t < 1.09$ (Lee et al., 2009). It has been found that the tolerance factor for ideal cubic perovskite structure is unity. If the tolerance factor is slightly greater than unity ($t > 1$), the structure of perovskite is stabilized in tetragonal or hexagonal symmetry. However, if the tolerance factor is smaller than one ($t < 1$), the structure is often considered as rhombohedral or monoclinic (Snel et al., 2005; Wood, 1951; Keith, 1954; Roth, 1957).

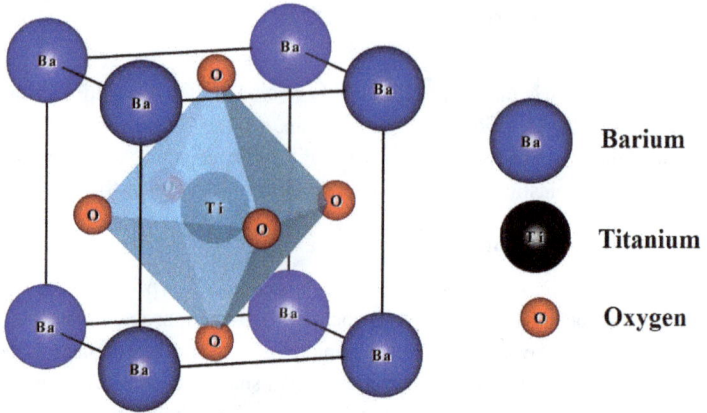

Figure 1.5 Unit cell of BaTiO₃ perovskite structure (ABO₃).

1.2.6 Phase transition

Phase transition is the transformation of the structure of a material from one thermodynamic phase to another phase. The crystal structure of many dielectric materials changes with temperature (i.e., they undergo a phase transition). The phase transition in crystals is due to the change in the forces of interaction between atoms in crystals. This change may produce various new properties in the crystal. The phase transition that produces or alters the spontaneous polarization is called ferroelectric phase transition. By changing the temperature or pressure, the atomic arrangements in the crystals may be changed without any change in chemical compositions (Damjanovic, 1998). The difference in crystal structure on either side of T_C may be large or small.

The relative permittivity of most ferroelectric materials can be changed considerably with changes in temperature (Moulson and Herbert, 2003). As the temperature changes, ferroelectric materials show dielectric anomalies with maxima at certain temperatures corresponding to crystallographic phase transition temperatures. The temperature corresponding to the maximum relative permittivity, at the ferroelectric to paraelectric transition is known as the Curie point (T_C). In the case of BaTiO₃, a transition from ferroelectric tetragonal phase to paraelectric cubic phase occurs above ~130°C (figure

1.6). The other ferroelectric phases are orthorhombic and rhombohedral at a temperature of ~ 0°C and ~ -90°C respectively.

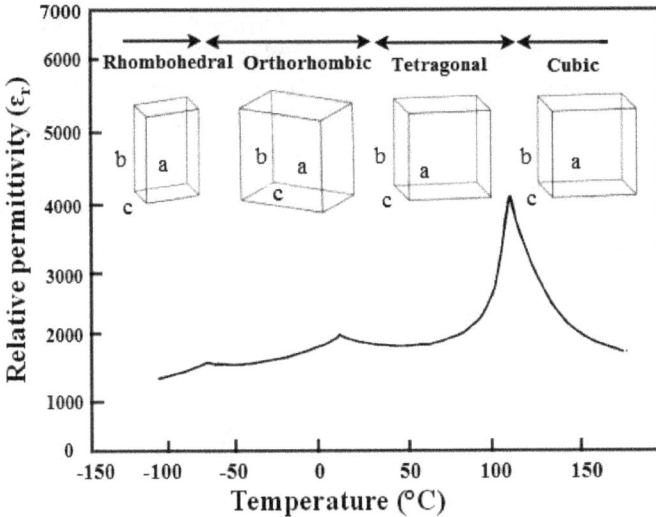

Figure 1.6 Variation of dielectric constant with temperature showing the phase transition in BaTiO₃, which undergoes four structural changes.

1.2.7 Morphotropic phase boundary (MPB)

A morphotropic phase boundary is defined as an abrupt change in the crystal structure of a solid solution with variation in composition. It is well known that the optimal piezoelectric response can be ascribed to the enhanced polarization orientation because of the co-existence of different phases between rhombohedral, tetragonal and/or orthorhombic ferroelectric phases (Ding, 2010). At such a phase boundary, two or more different structures with different polarizabilities with similar free energies may co-exist. The term morphotropic phase boundary (MPB) was proposed by Jaffe *et al.*, (1971) to describe the boundary between the two phases in the PZT phase diagram (Figure 1.7) (Jafee *et al.*, 1971). The poling process of $Pb(Zr_{1-x}Ti_x)O_3$ ceramic is also easy at the composition of morphotropic phase boundary because the spontaneous polarization

Lead-free Piezo-Ceramic Solid Solutions, R. Saravanan   Materials Research Forum LLC
Materials Research Foundations **41** (2018)   doi: http://dx.doi.org/10.21741/9781945291951

within each grain can be switched to one of the 14 possible orientations (eight [111] directions for the rhombohedral phase and six [100] directions for the tetragonal phase).

MPB is particularly interesting for both applications and fundamental studies because the useful properties such as piezoelectric coefficient (d_{33}), dielectric permittivity (ε) and remnant polarization (P_r) all reach a maximum value at the boundary. The highest piezoelectric response ($d_{33} \sim 350$ pC/N) in $Pb(Zr_{1-x}Ti_x)O_3$ system (x=0.47) was found near the boundary between tetragonal and rhombohedral phases (Guo *et al.*, 2000), as shown in figure 1.7. Recent experimental work by Noheda and co-worker suggested that a monoclinic phase exist at a composition very near the MPB initiating a significant number of experimental and theoretical works (Noheda and Gonzalo, 1973; Zachariasz and Bochenek, 2009).

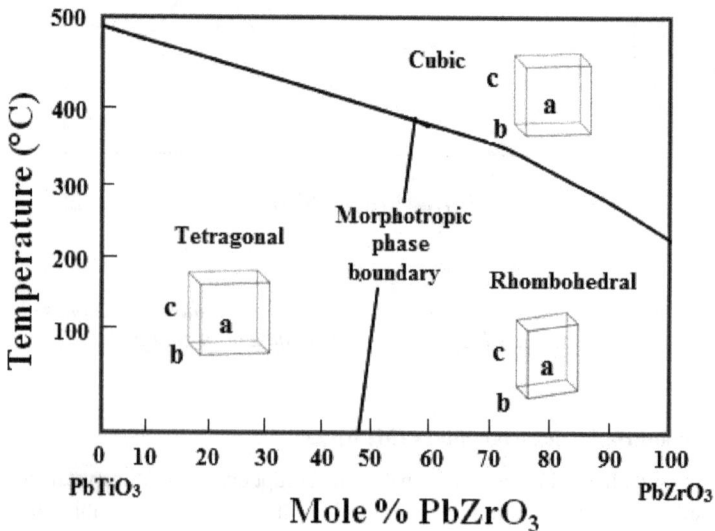

Figure 1.7 Morphotropic phase boundary in $Pb(Zr_{1-x}Ti_x)O_3$ system.

1.3   Review of Literature

1.3.1   Lead-based piezoelectric materials

1.3.1.1 Lead zirconate titanate (PZT)

The most widely used piezoelectric ceramics are lead-based materials. The solid solution of lead zirconate titanate (PZT) is one of the most frequently studied ferroelectric material, which is useful in actuating and sensing applications. The $Pb(Zr, Ti)O_3$ has the ABO_3 type perovskite structure with Ti^{4+} ions and Zr^{4+} ions occupying B-sites. The structure of PZT is cubic above the transition temperature (T_C) and the structure of PZT changes to ferroelectric or antiferroelectric phase below the transition temperature (T_C) (Noheda and Gonzalo, 1973). The tetragonal and the rhombohedral ferroelectric phases are divided by the morphotropic phase boundary (MPB) at a particular composition of $Pb(Zr_{1-x}Ti_x)O_3$ solid solution. There is an abrupt change in the lattice constants for $Pb(Zr_{1-x}Ti_x)O_3$ solid solution with Ti:Zr (52:48) ratio at room temperature, which is attributed to the morphotropic phase boundary (MPB). The enhanced electrical properties such as piezoelectric coefficients (d_{33}~300-550 pC/N), dielectric permittivity and remnant polarization can be observed in this region due to the co-existence of tetragonal and rhombohedral phases.

The anomalous properties near the MPB can be explained by a phase transition between the tetragonal and rhombohedral phases (Zhang et al., 1997; Sheen and Kim, 2003). Lead zirconate titanate ($Pb(Zr_{1-x}Ti_x)O_3$) is a solid solution of ferroelectric $PbTiO_3$ (T_C=490°C) and antiferroelectric $PbZrO_3$ (T_C=230°C). At room temperature, the compositions of PZT possess piezoelectric and pyroelectric properties for 0.042<x<0.380 (rhombohedral R3c), 0.380<x<0.470 (rhombohedral R3m) as well as 0.48<x<1.000 (tetragonal P4mm). PZT-based ceramics have high piezoelectric coefficients, low cost and ease of reproducibility (Zachariasz and Bochenek, 2009). Although, lead-based perovskite ceramics are dominating the markets of both the piezoelectric and the ferroelectric devices, the toxicity of lead has raised serious environmental issues. Hence, the growing demand for replacing these lead-based piezoelectrics has paid more focus on the development of efficient lead-free systems.

1.3.2   Lead-free piezoelectric materials

More than a decade of intense research provides primordial interest for developing lead-free ceramics as an alternative of lead-based materials in many technological applications.

Amongst the several systems investigated, $BaTiO_3$, $Na_{0.5}Bi_{0.5}TiO_3$ (NBT) and (Na, K)NbO_3 (NKN)-based systems are the most promising lead-free ceramics (Aksel *et al.*, 2010). Liu *et al.*, (2009) first reported a lead-free pseudo binary $Ba(Zr_{0.2}Ti_{0.8})O_3$-$x(Ba_{0.7}Ca_{0.3})TiO_3$ ferroelectric morphotropic phase boundary (MPB) systems to replace PZT-based systems, which show a high piezoelectric coefficient of d_{33}~620 pC/N at x=0.5 (Liu *et al.*, 2009; Bao *et al.*, 2010). Saito *et al.*, (2004) reported that a giant d_{33} of 400-413 pC/N could be obtained in Li^+, Ta^{5+} and Sb^{5+} modified NKN ceramics using the orthorhombic-tetragonal phase boundary. Hence, the construction of a new rhombohedral-tetragonal phase boundary has been used to enhance the piezoelectric properties of NKN-based ceramics (Saito *et al.*, 2004). To replace the lead-based piezoelectrics, a lot of studies related to lead-free piezoelectric materials have been conducted (Li *et al.*, 2004; Tian *et al.*, 2013).

1.3.3 Need for the lead-free piezoelectric ceramics

Piezoelectric materials are nowadays widely used as transducers, actuators or sensors due to their excellent piezoelectric properties. Piezoceramics are found in many applications as in ultrasonic transducers, mobile phones, etc. In general, lead can damage organ system of human and induce numerous disease symptoms like headaches, constipation, nausea, anemia, nerve, brain, kidney damage. For children, the exposure to high lead concentrations can cause slowed growth, behavioural disorders or learning disabilities (Takenaka *et al.*, 2005). It is well known that lead remains in the environment for a long time and it is therefore important to solve the problem of recycling of lead-containing devices, as for example computers. In that context, the most promising lead-free piezoelectric materials are barium titanate and sodium potassium niobate based compounds. Several other materials also possess promising electrical properties in the compositions near the phase transition.

1.3.4 Structural and piezoelectric properties of lead-free materials

1.3.4.1 Barium titanate (BaTiO₃)

Barium titanate is one of the known fundamental lead-free ferroelectric material and has been used in electrical components such as multilayered ceramic capacitors and positive temperature coefficient thermistors. Barium titanate has high permittivity at room temperature (ε~1000-5000) (Zhao *et al.*, 2004; Moulson and Herbet, 2003). Barium titanate has a perovskite structure with Ba^{2+} on the A-site and Ti^{4+} on the B-site. Barium titanate exhibits phase transition with various temperatures (T). At T $\leq$ -90°C, the structure of $BaTiO_3$ is trigonal or rhombohedral, between -90°C to 5°C, the rhombohedral structure changes to orthorhombic. Above 5°C, the orthorhombic structure changes to tetragonal and finally the structure changes to cubic at 120°C (Curie temperature (T_C)) (Maiti *et al.*, 2008). At each phase boundary of two phase transitions, the electric properties including dielectric permittivity, capacitance and piezoelectric operation modes have been enhanced.

1.3.4.2 $Na_{0.5}K_{0.5}NbO_3$-$BaTiO_3$ based solid solutions (NKN-BT)

NKN-BT-based solid solution system consists of three morphotropic phase boundaries at room temperature. The phase transitions are orthorhombic to tetragonal in the range $0.0 \leq x \leq 0.1$, tetragonal to cubic in the range $0.1 \leq x \leq 0.2$. Among these MPBs, rhombohedral to tetragonal transition in the range $0.0 \leq x \leq 0.1$ gives best piezoelectric properties (Ahn *et al.*, 2008; Choi *et al.*, 2007). It is very difficult to obtain pure $Na_{0.5}K_{0.5}NbO_3$ (NKN) ceramics by ordinary sintering process because of the high volatility of Na_2O and K_2O compounds at high temperatures. To improve densification and piezoelectric properties of NKN ceramics, some dopants are added into NKN ceramics to form a solid solution, such as NKN-$BaTiO_3$ (Ahn *et al.*, 2007), NKN-$LiNbO_3$ (Guo *et al.*, 2004) and NKN-$LiSbO_3$ (Zhang *et al.*, 2006). Effect of addition of small amounts of $BaTiO_3$ ($0.0 \leq x \leq 0.20$) has also been studied on $(K_{0.5}Na_{0.5})_{0.96}Li_{0.04}NbO_3$ (Ahn *et al.*, 2008). Saito *et al.*, (2004) reported that NKN-based $(K_{0.44}Na_{0.52}Li_{0.04})(Nb_{0.84}Ta_{0.10}Sb_{0.06})O_3$ ceramics show higher piezoelectric coefficient (d_{33}~300 pC/N) (Saito *et al.*, 2004).

1.3.4.3 $Na_{0.5}Bi_{0.5}TiO_3$-$BaTiO_3$ based solid solutions (NBT-BT)

$BaTiO_3$ (BT) and $Na_{0.5}Bi_{0.5}TiO_3$ (NBT) are two interesting and attractive lead-free piezoceramics, which are considered to be the good candidates to replace the widely used lead-containing perovskite materials. NBT is a ferroelectric material which has relatively high Curie temperature (T_C=320°C), a relative large remnant polarization (P_r=38 μC/cm²) and a coercive field of E_C=73 kV/cm at room temperature (Smolenskii *et al.*, 1961). But, pure NBT is hard to be poled due to its large coercive field and relatively large conductivity. The solid solution of $Na_{0.5}Bi_{0.5}TiO_3$ with $BaTiO_3$ has been investigated to enhance the piezoelectric properties (Smolenskii *et al.*, 1961). Among the NBT-BT based systems, $(1-x)(Na_{0.5}Bi_{0.5})TiO_3$-$xBaTiO_3$ ceramics possess a rhombohedral-tetragonal morphotropic phase boundary (MPB) between x=0.06-0.07, where the system shows outstanding piezoelectric and dielectric properties. Therefore, many research works focused on compositions near the MPB (Takenaka *et al.*, 1991). Lidijici *et al.*, (2015) reported that the (1-x)NBT-xBT ceramics with various compositions show a morphotropic phase boundary (MPB) at the composition x, between 0.065 and 0.07

(Lidijici *et al.*, 2015). At this composition, the system exhibits improved piezoelectric constant (d_{33}=150 pC/N).

1.3.4.4 $K_{0.5}Bi_{0.5}TiO_3$-$BaTiO_3$ based solid solutions (KBT-BT)

Potassium bismuth titanate, $K_{0.5}B_{0.5}TiO_3$ (KBT) is a ferroelectric material synthesized first by Popper *et al.* in 1957 (Popper *et al.*, 1957, Rödel *et al.*, 2009). KBT is a typical ABO_3 type material with K^+ and Bi^{3+} on A-site and Ti^{4+} on B-site (Smolenskii *et al.*, 1961). X-ray analysis shows that the crystal structure of KBT is tetragonal with space group P4mm at room temperature with c/a ratio ~1.018. The structure of KBT transforms to a pseudocubic structure around 270°C and becomes completely cubic (Pm3m) around 410°C (Ivanova *et al.*, 1962). The dielectric profiles of KBT show a broad peak in ε_r-T plots with phase transition temperature at T_C~380°C (Hiruma *et al.*, 2005; Otoničar *et al.*, 2010). One of the main challenges of this system is that it is difficult to produce dense ceramics using ordinary firing methods. This low density makes the materials difficult to fully pole. The density can be improved through the use of sintering aids with the addition of $BaTiO_3$ in the $K_{0.5}B_{0.5}TiO_3$ ceramics. With the addition of 10 mol% $BaTiO_3$, dense textured materials were formed using ordinary sintering methods (Nemoto *et al.*, 2008). Hiruma *et al.*, (2007) found that processing of KBT with excess Bi_2O_3 improved the piezoelectric and ferroelectric properties of the material (d_{33}=101 pC/N) because Bi_2O_3 acts as a sintering aid and prevents the formation of micro-cracks (Hiruma *et al.*, 2007).

1.4   Methodologies used for analysis

1.4.1   Structural determination using Rietveld refinement technique

To understand the properties of materials, it is essential to know about their atomic structure. The X-ray diffraction methods are the best techniques to study about the atomic structure of the materials. In the past three decades, powder diffraction has played a vital role in the field of structural physics, chemistry and material science. High temperature superconductors and high pressure research have relied mostly on power diffraction techniques. In early 1990s, powder diffraction data have been used for the structural determination of only very few crystals. But today, numerous organic and inorganic crystal structures have been solved through powder diffraction data. Many powder diffraction methodologies have developed which have contributed to analyze the structure of the compounds. Since the powder diffraction peaks grossly overlap in the conventional powder diffraction method, it prevents the exact determination of the crystal structure. The Rietveld method (Rietveld, 1969) minimizes the impact of these overlapping peaks and determines the real crystal structure. Rietveld method (Rietveld, 1969) is a technique for the crystal structure refinement which uses the whole powder diffraction pattern, devised by Hugo Rietveld. In Rietveld method (Rietveld, 1969), least-squares approach is used so that the calculated and measured diffraction profiles are optimized and by iterative technique, the profiles are refined. The following sections explain the powder profile parameters that we come across in Rietveld (Rietveld, 1969) refinement technique.

1.4.1.1 Rietveld method

The Rietveld (Rietveld, 1969) analysis is a unique and valuable method for extracting detailed structural information from X-ray powder data. Rietveld method (Rietveld, 1969) was devised to quantify the structural treatment, doping, compression and strain. This refinement can be performed on the obtained experimental powder X-ray diffraction data. The method was introduced by Hugo Rietveld in 1967 (Rietveld, 1969). It is a whole profile structure refinement method based on least-square fitting. A calculated profile fit of the structure is matched with experimental data and then the selected parameters are refined repeatedly in order to get the best fit between the calculated and experimental profiles.

The Rietveld method (Rietveld, 1969) is a technique for refining structural parameters and lattice parameters directly from whole powder diffraction data sets. The structural parameters are fractional coordinates, occupation factors, isotropic/anisotropic atomic displacement parameters, etc. This method was initially only applied to neutron data due to their simple peak shape. The Rietveld refinement (Rietveld, 1969) uses different peak-shape functions such as Lorentzian and pseudo-Voigt. Then, the application of this method was extended to laboratory X-ray diffractometers by Malmros and Thomas in 1977 (Malmros and Thomas, 1977). It should be mentioned that successful structural analysis by Rietveld method (Rietveld, 1969) is directly related to the quality of powder diffraction data. The structural refinement gives model of the crystal structure. This model has approximations on unit cell dimensions and atomic coordinates and it should have the same space group. The refinement will be satisfactory when the best fit between the observed and the calculated model is achieved. The Rietveld method (Rietveld, 1969) uses a least-squares approach, to minimise the weighted sum of the point-by-point differences squared in a powder diffraction pattern. The intensity is calculated using the experimental parameters such as instrumental parameters, background intensities, absorption, extinction etc. and the sample dependent parameters such as unit cell parameters, atomic fractional coordinates, atomic occupancy of each crystallographic site, Debye-Waller factors and background. Minimization (M) is defined by an equation such as,

$$M = \sum_i w_i \left(y_{i\,(obs)} - y_{i\,(calc)}\right)^2 \tag{1.5}$$

w_i is weighting factor for each observed point and is equal to $w_i = 1/y_i$. $y_{i\,(obs)}$ and $y_{i\,(calc)}$ are observed and model calculated intensities at each step for i data points. The computer program for structural analysis and refinement used in this thesis is JANA 2006 (Petříček *et al.*, 2006). Constraints can be applied to reduce the degree of freedom of the set of equations that has to be solved. The quality of the Rietveld refinement (Rietveld, 1969) is indicated by some residual functions. The profile R-factor, which is the most straightforward discrepancy index, is a measure of the disagreement between the observed and calculated profile. The quality of the fit can be monitored by the R_p (profile factor), and R_{wp} (weighted profile factor), the goodness of fit is indicated by the χ^2 value, which is optimal when $\chi^2 = 1$, representing an ideal match between experimental and theoretical data, which is given as:

$$R_p = \frac{\sum y_{i(obs)} - y_{i(cal)}}{\sum y_{i(obs)}} \tag{1.6}$$

$$R_{wp} = \frac{\sum w_i (y_{i(obs)} - y_{i(cal)})^2}{\sum w_i (y_{i(obs)})^2} \tag{1.7}$$

$$\chi^2 = \frac{\sum w_i (y_{i(obs)} - y_{i(cal)})^2}{N - P} \tag{1.8}$$

N - number of data points
P - number of refined parameters

1.4.2 Electron density distribution

The quantum mechanical theory explains that the electron density is the measure of the probability of an electron being present at a specific location. The atoms are surrounded by electron clouds. The electron density is defined as the number of electrons per unit volume. The chemical bondings as well as the physical and chemical properties of the crystal systems have been analyzed by the electron density of a system. The electron density distribution study is applied in many disciplines in chemistry, physics, biology and geology (Stout and Jensen, 1970). The study of chemical bonding and internal local structure of a crystalline system is very important and it gives useful information about the transport properties which can be effectively utilized for device applications. For the precise understanding of the nature of chemical bonds, it is essential to study about the electron density distribution between the atoms.

Since the lattice has a periodicity, the electron density is also considered to behave as a periodic function. The number of electrons in any volume element dV is $\rho(x,y,z)dV$. In a X-ray scattering experiment, the wavelet scattered by this element is

$$\rho(x,y,z)\exp\left[-2\pi i(hx + ky + lz)\right]dV \tag{1.9}$$

The resultant sum of contributions from all the elements in the unit cell, i.e., the integral over its volume gives

$$F_{hkl} = \int \rho(x,y,z) \exp[-2\pi i(hx + ky + lz)]dV \tag{1.10}$$

The structure factor is considered as a resultant of adding the scattered waves in the direction of the *hkl* reflection from the atoms in the unit cell. This approach was based on the assumption that the scattering power of the electron cloud surrounding each atom could be equated to that of the proper number of electrons concentrated at the atomic centre. But the structure factor may equally well be considered as the sum of the wavelets scattered from all the infinitesimal elements of electron density in a unit cell, with no assumptions being made about the distribution of this density. The electron density $\rho(r)$ is defined as the number of electrons per unit volume.

The geometric properties of unit cells can be deduced from the locations of reflections on various kinds of X-ray diffraction photographs. It concerned with the measurement of the relative intensities of these reflection, since it is from the intensities that it hopes to be able to deduce the electron density distribution in the crystal cell. There are connections between the intensities and the electron density distribution. There are a few general precautions, which are applicable to any intensity measuring method. If the structure factors and phases are known, the electron-density distribution of the unit cell can be calculated. The magnitude of individual structure factors are calculated as the square-root of the measured diffraction intensity and their phases are determined by solving the structure. The interpretation is described as a model, which is improved by least-squares refinement based on the structure factors. The electron density can then be calculated as a Fourier summation of phased structure factors.

Lead-free Piezo-Ceramic Solid Solutions, R. Saravanan | Materials Research Forum LLC
Materials Research Foundations **41** (2018) | doi: http://dx.doi.org/10.21741/9781945291951

Intensities of diffracted X-rays are due to interference effects of X-rays scattered by all the different atoms in the structure. The diffraction pattern is the Fourier transform of the crystal structure, corresponding to the pattern of waves scattered from an incident X-ray beam by a single crystal; it can be measured by experiment (only partially, because the amplitudes are obtainable from the directly measured intensities via a number of correction, but the relative phases of the scattered waves are lost), and it can be calculated (giving both amplitudes and phases) for a known structure. So, the crystal structure is the Fourier transform of the diffraction pattern and is expressed in terms of electron density distribution concentrated in atoms; it cannot be measured by direct experiment, because the scattered X-rays cannot be refracted by lenses to form an image as done with light in an optical microscope, and it cannot be obtained directly by calculation, because the required relative phases of the waves are unknown. We can calculate the electron density distribution given a set of structure factors, using the Fourier series.

1.4.2.1 Fourier method

Any well-behaved function can be represented by means of suitable series of trigonometric terms called Fourier series. We can imagine the unit cell is divided into small volumes dV in which there are $\rho(r)\,dV$ number of electrons. The scattered amplitude from such a small volume will be $\rho(r)\,dV$ times as much that from an electron at the same position. From this we find the total scattered amplitude from the distribution of electron density $\rho(r)$.

$F(H)$ can be expressed in terms of density $\rho(r)$ as

$$F(H) = \int \rho(r)\exp[2\pi i H \cdot r]dV \tag{1.11}$$

The inverse Fourier transform of this gives the electron density

$$\rho(r) = \int \rho(x,y,z)\exp[-2\pi i H \cdot r]dV = \frac{1}{V}\sum F(H)\exp(-2\pi i H \cdot r) \tag{1.12}$$

Since $F(H)$ is defined at the discrete set of reciprocal lattice points k, the integral is replaced by the summation. Writing the structure factor as $F(H) = A(H) + iB(H)$, then

$$\rho(r) = \frac{1}{V}\sum [(A(H) + iB(H))[\cos(2\pi H \cdot r) - i\sin(2\pi H \cdot r)] \tag{1.13}$$

where A(H) = A(-H) and B(H) =B(-H) since the electron density is a real function.

Therefore,

$$\rho(r) = \frac{1}{V}\Sigma_{1/2}[2A(H)cos(2\pi H \cdot r) + 2Bsin(2\pi H \cdot r)] \qquad (1.14)$$

with A(H) = $|F(H)|cos\varphi$ and B(H) = $|F(H)|sin\varphi$ resulting in

$$\rho(r) = \frac{2}{V}\Sigma_{1/2}[|F(H)|cos\varphi cos(2\pi H \cdot r) + |F(H)|sin\varphi \, sin(2\pi H \cdot r)] \qquad (1.15)$$

which reduces to

$$\rho(r) = \frac{2}{V}\Sigma_{1/2}[|F(H)|cos(2\pi H \cdot r) - \varphi(H)] \qquad (1.16)$$

In other words, each structure factor contributes a plane wave to the total density with wave vector H and phase φ. As we know the formation of the image, which is the density, requires knowledge of the phases of the structure factors. Once an approximation to the scattering density is known, $\varphi(H)$may be calculated on the basis of this approximation, and an admittedly imperfect image of the structure can be obtained. At the same time, anomalous scattering can be corrected for which can be done by subtracting the calculated contributions $\Delta A_{calc}^{anamalous}$ and $\Delta B_{calc}^{anamalous}$ from A and B, respectively, using the anomalous scattering factors and f' and f''.

The period of the plane wave with amplitude F(H), in the direction of the wave vector H, equals 1/H. The period is therefore shorter for higher-order reflections are included in the summation, the resolution of the image improves. The improvement is analogous to the increase in resolution in an optical image obtained with shorter-wavelength radiation.

The non-existence of lenses for X-ray beams makes it necessary to use computational methods to achieve the Fourier transform of the diffraction pattern into the image. In the calculation of charge density by this method we need infinite number of Fourier co-efficient to perform the Fourier Synthesis. But we use only a limited number of Fourier co-efficient and ignore experimental errors by setting all the missing Fourier co-efficient as zero. We neglect the missing structure factors by setting them to zero simply because the experiment cannot be or was not carried out. This is a highly biased assumption. This results in the unphysical negative electron density and hampers the use of it in understanding the finite details like the bonding charge in valence region.

1.4.2.2 Maximum entropy method

Experimental electron density distribution can be reconstructed from accurate X-ray diffraction data and analysis steps (Iversen *et al.*, 1996). Maximum entropy method (MEM) (Collins, 1982) is used to reconstruct the charge density distributions in the molecular systems. To understand the physical and chemical properties of the material systems, one should require the knowledge about their charge distribution (Bader, 1991). The resultant density distribution gives detailed information about the structure, without using the structural model. MEM (Collins, 1982) electron density map is an accurate mathematical tool for structural analysis. Compared to the map drawn by conventional Fourier transformation, the resolution of MEM electron density map is higher (Sakata and Sato, 1990). MEM uses the structure factors retrieved from Rietveld (Rietveld, 1969) refinement. Hence, the combination of MEM (Collins, 1982) and Rietveld (Rietveld, 1969) method provide the detailed structure model. Using XRD, it is impossible to collect the exact values of all the structure factors. Hence, there are some errors in the number of observed structure factors by the experiment. The uncertainties due to these errors must be rectified. MEM (Collins, 1982) introduces the concept of entropy to tackle the uncertainty properly. So, the concept behind maximum entropy method (MEM) (Collins, 1982) is to obtain the electron density distribution which is consistent with the observed structure factors and to leave the uncertainties minimum. The mathematical description of MEM (Collins, 1982) is explained below.

The maximum entropy method (MEM) is an information-theory-based technique to enhance the information was developed in the field of radio astronomy to enhance the information obtained from noisy data (Gull and Daniel, 1978). The theory is based on the same equations that are the foundation of statistical thermodynamics. Both the statistical entropy and the information entropy deal with the most probable distribution. In the case of statistical thermodynamics, this is the distribution of the particles over position and momentum space, while in the case of information theory, the distribution of numerical quantities over the ensemble of pixels is considered. The probability of a distribution of N identical particles over m boxes, each populated by n_i particles, is given by

$$P = \frac{N!}{n_1! n_2! n_3! \ldots n_m!} \tag{1.17}$$

As in statistical thermodynamics, the entropy is defined as ln P. Since the numerator is constant, the entropy is, apart from a constant, equal to

$$S = -\sum_i n_i \ln n_i \tag{1.18}$$

where Stirlings' formula ((ln N! =N ln N - N) has been used.

In case there is a prior probability q_i for box i to contain n_i particles, then, becomes

$$P = \frac{N!}{n_1! n_2! n_3! \ldots n_m!} \times q_1^{n_1} q_2^{n_2} \ldots \ldots q_m^{n_m} \tag{1.18}$$

which gives, for the entropy expression,

$$S = -\sum_i n_i \ln n_i + \sum_i n_i \ln q_i = -\sum_{i=1}^m n_i \ln \frac{n_i}{q_i} \tag{1.19}$$

The maximum entropy method was first introduced into crystallography by Collins (Collins, 1982), who based on equation (1.19), expressed the information entropy of the electron density distribution as a sum over M grid points in the unit cell, using the entropy formula (Jaynes, 1968).

$$S = -\sum_r \rho'(r) \ln \left(\frac{\rho'(r)}{\tau'(r)} \right) \tag{1.20}$$

where both $\rho'(r)$ and prior probability $\tau'(r)$ are related to the actual electron density in a unit cell as

$$\rho'(r) = \frac{\rho(r)}{\sum_r \rho(r)} \quad \text{and} \quad \tau'(r) = \frac{\tau(r)}{\sum_r \tau(r)} \tag{1.21}$$

where $\rho(r)$ and $\tau(r)$ are the electron density and prior electron density at a fixed r in a unit cell respectively. In the present theory, the actual densities are treated instead of normalized densities and $\rho'(r)$ become $\tau'(r)$ when there is no information. The $\rho'(r)$ and $\tau'(r)$ are normalized as

$$\sum \rho'(r) = 1 \quad \text{and} \quad \sum \tau'(r) = 1 \tag{1.22}$$

Materials Research Forum LLC
doi: http://dx.doi.org/10.21741/9781945291951

The entropy is maximized subject to the constraint

$$C = \frac{1}{N} \sum_k \frac{|F_{cal}(k) - F_{obs}(k)|^2}{\sigma^2(K)} \tag{1.23}$$

Where N is the number of reflections used for MEM analysis, $\sigma(k)$, standard deviation of $F_{obs}(k)$, the observed structure factor and $F_{cal}(k)$ is the calculated structure factor given by

$$F_{cal}(k) = V \sum_r \rho(r) exp(-2\pi i k. r) \, dV \tag{1.24}$$

where V is the volume of the unit cell. The constraint C is known as weak constraint, in which the calculated structure factors agree with the observed structure factors as a whole when C becomes unity. Equation (1.24) shows, the structure factors are given by the Fourier transform of the electron density distribution in a unit cell. In the MEM (Collins, 1982) analysis, there is no need to introduce the atomic factors, by which the structure factors are normally written. It should be emphasized here that it would be an assumption to use the atomic form factors in the formulation of the structure factors. Equation (1.24) guarantees that it is possible to introduce any kind of deformation of the electron densities in real space as long as information concerning such a deformation is included in the observed data.

We use Lagrange's method of undetermined multiplier (λ) in order to constrain the function C to be unity while maximizing the entropy. We then have

$$Q = S - \frac{\lambda}{2} C \tag{1.25}$$

$$Q = -\sum \rho'(r) ln\left(\frac{\rho'(r)}{\tau(r)}\right) - \frac{\lambda}{2N} \sum_k \frac{|F_{cal}(k) - F_{obs}(k)|^2}{\sigma^2(k)} \tag{1.26}$$

And when $dQ/d\rho = 0$ and using the approximation,
$ln\, x = x - 1$ we get,

$$\rho(r_i) = \tau(r_i) exp\left\{\left(\frac{\lambda F_{000}}{N}\right) \left[\sum \frac{1}{\sigma(H)^2}\right] |F_{obs}(k) - F_{cal}(k)| exp(-2\pi j\, k. r)\right\} \tag{1.27}$$

Materials Research Forum LLC
doi: http://dx.doi.org/10.21741/9781945291951

where $F_{000} = Z$ is the total number of electrons in a unit cell. In order to solve equation (1.28) an approximation is introduced to replaces $F_{cal}(\boldsymbol{k})$ as

$$F_{cal}(\boldsymbol{k}) = V \sum \tau(r) exp(-2\pi i\, \boldsymbol{k}.\boldsymbol{r})\, dV \qquad (1.28)$$

This approximation can be called zeroeth order single pixel approximation. By using this approximation the right hand side of equation (1.45) becomes independent of $\tau(\boldsymbol{r})$ and equation (1.28) can be solved in an iterative way starting from a given initial density for the prior distribution. A uniform density distribution is given as the prior density.

$$\tau(\boldsymbol{r}) \leq \tau(\boldsymbol{r}) \geq \frac{Z}{M} \qquad (1.29)$$

where M is the total number of pixels for which the electron density is calculated. The reason for this choice of prior distribution is that uniform density distribution corresponds to the maximum entropy state among all possible density distributions. In the calculation of $\rho(\boldsymbol{r})$, all of the symmetry recruitments are satisfied and the number of electrons (Z) is always kept constant through an iteration process. Mathematically, the summation concerning $\rho(\boldsymbol{r})$ in the above equations should be written as an integral. Since we must use a very limited number of pixels in the numerical calculation, the integral is replaced by the summation in the above equations (equations 1.18 to 1.29).

After completion of the MEM (Collins, 1982) enhancement, it becomes possible to evaluate the reflections missing from the summation. In a Fourier summation, the amplitudes of the unobserved reflections are assumed to be equal to zero, while the MEM (Collins, 1982) technique provides the most probable values. When extinction is present in the data set, it must be corrected for before the MEM procedure is started. The structure factors must similarly be corrected for anomalous scattering, if present. Both corrections require a model for their evaluation. The independent-atom model is usually adequate for this purpose. The advantage of maximum entropy method is a statistical deduction that can yield a high resolution density distribution from a limited number of diffraction data without using a structural model. It has been suggested that MEM (Collins, 1982) would be a suitable method for examining electron densities in the inner atomic region, for example, bonding region. It gives less biased information on the electron densities as compared to conventional Fourier synthesis.

1.4.2.3 Methodology for the determination of charge density

The technological advances in recent years bring demands for integrated three dimensional visualization systems to deal with both structural models and volumetric data, such as electron and nuclear densities. The crystal structures and spatial distribution of various physical quantities obtained experimentally and by computer simulations should be understood three-dimensionally. Once the structure factors are refined, they are further utilized for the evaluation of MEM (Collins, 1982) charge density. The maximum entropy method (Collins, 1982) gives information on reconstructing the structure factor using preliminary information like position, type, space group, etc. The calculated structure factor is then compared with the observed one and the resultant calculated structure factor and observed one, which is used for the reconstruction of charge density using MEM analysis (Collins, 1982). The electron density distributions in the unit cell are constructed through the PRIMA (Practice of Iterative MEM Analyses) (Izumi and Dilanan, 2002) software. PRIMA (Izumi and Dilanan, 2002) is a program to calculate electron densities from X-ray diffraction data. The input file contains the cell parameters, space group, pixels, total charge, Lagrange parameter and structure factors.

In the present work, the unit cell was divided into $64\times64\times64$ pixels and the initial electron density at each pixel was fixed uniformly as Z/a_0^3, where Z is the number of electrons in the unit cell. The electron density is evaluated by carefully selecting the Lagrange multiplier in each case such that the convergence criterion C becomes unity after performing minimum number of iterations. The three dimensional (3D) electron density was plotted using VESTA (Visualization of Electronic and Structural Analysis) (Momma and Izumi, 2006) software package. VESTA (Momma and Izumi, 2006) software deals with structural models and volumetric data at the same window. To understand the nature of bond in the materials, two dimensional (2D) and one dimensional (1D) distribution of electron densities on different lattice plane have been mapped and discussed in the following chapters.

References

[1] Ahn C.W., Choi C.H., Park H.Y., Nahm S., J. Mater. Sci. Lett. 43(20), 6784-6797 (2008). https://doi.org/10.1007/s10853-008-2934-1

[2] Ahn C.W., Park C.S., Viehland D., Nahm S., Kang D.H., Bae K.S., Priya S., Jpn. J. Appl. Phys. 47, 8880-8883 (2008). https://doi.org/10.1143/JJAP.47.8880

[3] Ahn C.W., Park H.Y., Nahm S., Uchino K., Lee H.G., Lee H.J., Sensor. Actuator. A-Phys. 136, 255-260 (2007).

[4]   Aksel E., Jones J.L., Sensors. 10, 1935-1954 (2010).
        https://doi.org/10.3390/s100301935

[5]   Bader R.F.W., Atoms in molecules, A Quantum Theory, Oxford University Press,
        (1991).

[6]   Bao H., Zhou C., Xue D., Gao J., Ren X., J. Phys. D. 43, 465401 (2010).
        https://doi.org/10.1088/0022-3727/43/46/465401

[7]   Choi C.-H., Ahn C.W., Nahm S., Appl. Phys. Lett. 90(13), 132905 (2007).
        https://doi.org/10.1063/1.2717559

[8]   Collins D.M., Nature. 298, 49 (1982). https://doi.org/10.1038/298049a0

[9]   Damjanovic D., Reports on Progress in Physics. 61, 1267-1324 (1998).
        https://doi.org/10.1088/0034-4885/61/9/002

[10]   Ding A., W.H., Journal of Ceramic Processing Research. 11, 44-46 (2010).

[11]   Gene H.H., Ferroelectric ceramics: history and technology. J. Am. Ceram. Soc.
        82(4), 797-818 (1999). https://doi.org/10.1111/j.1151-2916.1999.tb01840.x

[12]   Gull S.F., Daniell G.J., Nature. 272, 686 (1978). https://doi.org/10.1038/272686a0

[13]   Goldschmidt V.M., Naturwissenschaften 14, 477 (1926).
        https://doi.org/10.1007/BF01507527

[14]   Guo R., Cross L.E., Park S.-E., Noheda B., Cox D.E., en G. Shirane G., Phys. Rev.
        Lett. 84(23), 5423-5426 (2000). https://doi.org/10.1103/PhysRevLett.84.5423

[15]   Guo Y., Kakimoto K., Ohsato H., Appl. Phys. Lett. 85(18), 4121-4123 (2004).
        https://doi.org/10.1063/1.1813636

[16]   Hiruma Y., Rintaro A., Nagata H., Takenaka T., Jpn. J. Appl. Phys. 44(7A) 5040-
        5044 (2005). https://doi.org/10.1143/JJAP.44.5040

[17]   Hiruma Y., Nagata H., Takenaka T., Jpn. J. Appl. Phys. Part 1, 46, 1081-1084
        (2007). https://doi.org/10.1143/JJAP.46.1081

[18]   Ivanova V.V., Kapyshev A.G., Venevtsev Y.N., Zhdanov G.S., Izv. Akad. Nauk
        SSSR. 26, 354-356 (1962).

[19]   Iversen B.B., Larsen F.K., Figgis B.N., Reynolds P.A., Acta Cryst. B53, 923-932
        (1996). https://doi.org/10.1107/S010876819600794X

[20]   Jaffe B., Cook W.R., Jeffe H., Piezoelectric ceramics. Academic press, London
        and New York, Vol. 3 (1971).

[21] Jaynes E.T., IEEE Trans. Syst. Sci. Cybern., SSC-4, 227 (1968).
https://doi.org/10.1109/TSSC.1968.300117

[22] Keith M.L. and Roy R., American Mineralogist. 39, 1-23 (1954).

[23] Lee W.-C., Huang C.-Y., Tsao L.-K., Wu Y.-C., Journal of the European Ceramic
Society. 29(8), 1443-1448 (2009).
https://doi.org/10.1016/j.jeurceramsoc.2008.08.028

[24] Li H.D., Feng C.D., Yao W.L., Mater. Lett. 58, 1194-1198 (2004).
https://doi.org/10.1016/j.matlet.2003.08.034

[25] Liu W., Ren X., Phys. Rev. Lett. 103(25) 1-4 (2009).
https://doi.org/10.1103/PhysRevLett.103.257602

[26] Lidijici H., Laghoun B., Berrahal M., Rguitti M., Hentatti M.A., Khemakhem H.,
Journal of Alloys and Compounds. 618, 643-648 (2015).
https://doi.org/10.1016/j.jallcom.2014.08.161

[27] Izumi F., Dilanian R.A., Recent Research Developments in Physics, Transworld
Research Network, Trivandrum. Vol. 3, Part II, 699-726 (2002).

[28] Malmros G., Thomas J.O., J. Appl. Cryst. 10, 7-11 (1977).
https://doi.org/10.1107/S0021889877012680

[29] Momma K., Izumi F., Commission on Crystallogr. Comput IUCr Newslett. 7, 106
(2006).

[30] Moulson A.J., Herbert J.M., Electroceramics: materials, properties and
applications, 2nd Edition, Wiley, New York (2003).
https://doi.org/10.1002/0470867965

[31] Maiti T., Guo R., Bhalla A.S., J. Am. Ceram. Soc., 91, 1769-80, (2008).
https://doi.org/10.1111/j.1551-2916.2008.02442.x

[32] Noheda B., Gonzalo J.A., The monoclinic phase in PZT: new light on
morphotropic phase boundaries, Physics department, Brookhaven National Lab.,
Upton, NY, 1, (1973).

[33] Nemoto M., Hiruma Y., Nagata H., Takenaka T., Jpn. J. Appl. Phys., 47(5), 3829-
3832 (2008). https://doi.org/10.1143/JJAP.47.3829

[34] Otoničar M., Škapin S.D., Jančar B., Ubic R., Suvorov D., Journal of the
American Ceramic Society, 93(12), 4168-4173 (2010).
https://doi.org/10.1111/j.1551-2916.2010.04013.x

[35]   Petříček V., Dušek M., Palatinus L., JANA 2006, The crystallographic computing system Institute of Physics Academy of sciences of the Czech republic, Praha (2006).

[36]   Popper P., Ruddlesden S., Ingles T., Trans. Br. Ceram. Soc. 56, 9 (1957).

[37]   Richerson D.W., Modern ceramic engineering: Properties, processing, and use in design, Second Edition, Marcel Dekker, inc., (1992).

[38]   Rietveld H.M., J. Appl. Crystallogr., 2, 65 (1969). https://doi.org/10.1107/S0021889869006558

[39]   Roth R.S., Journal of Research of the National Bureau of Standards. 58, 75-88 (1957). https://doi.org/10.6028/jres.058.010

[40]   Rödel J., Jo W., Seifert K.T.P., Anton E.-M., Granzow T., Damjanovic D., Journal of the American Ceramic Society, 92(6), 1153-1177 (2009). https://doi.org/10.1111/j.1551-2916.2009.03061.x

[41]   Saito Y., Takao H., Tani T., Nonoyama T., Takatori K., Homma T., Nagaya T., Nakamura M., Nature. 432, 84-87 (2004). https://doi.org/10.1038/nature03028

[42]   Sakata M., Sato M., Acta Cryst. A. 46, 263 (1990). https://doi.org/10.1107/S0108767389012377

[43]   Segal D., Chemical synthesis of advanced ceramic materials, Cambridge University Press, (1991).

[44]   Shannon R.D., Acta Cryst. A32, 751-767 (1976). https://doi.org/10.1107/S0567739476001551

[45]   Sheen D., Kim J., Phys. Rev. B 67, 144102 (2003). https://doi.org/10.1103/PhysRevB.67.144102

[46]   Spaldin N.A., Analogies and Differences between Ferroelectrics and Ferromagnets, in Physics of Ferroelectrics: A Modern Perspective, Berlin, Heidelberg: Springer Berlin Heidelberg, 175-218 (2007). https://doi.org/10.1007/978-3-540-34591-6_5

[47]   Smolenskii G.A., Isupov V.A., Agranovskaya A.I., Krainik N.N., Sov. Phys. Solid State. 2(11) 2651-2654 (1961).

[48]   Snel M., Groen W., de With G., Journal of the European Ceramic Society, 25(13), pp. 3229-3233 (2005). https://doi.org/10.1016/j.jeurceramsoc.2004.07.033

[49]   Stout G.H., Jensen L.H., X-ray structure determination, chapter 1, 2nd Edition, Wiley- Interscience publication, (1989).

[50]   Takenaka T., Maruyama K.-I., Sakata K., Jpn. J. Appl. Phys. 30(9B), 2236-2239 (1991). https://doi.org/10.1143/JJAP.30.2236

[51]   Takenaka T., Nagat H., J. Eur. Ceram. Soc. 25, 2693 (2005). https://doi.org/10.1016/j.jeurceramsoc.2005.03.125

[52]   Tian Y., Wei L., Chao X., Liu Z., Yang Z., J. Am. Ceram. Soc. 96, 496-502 (2013).

[53]   Wood D.L., Tauc J., Phys. Rev. B. 5, 3144 (1972). https://doi.org/10.1103/PhysRevB.5.3144

[54]   Wood E.A., Acta Crystallographica. 4, 353-361 (1951). https://doi.org/10.1107/S0365110X51001112

[55]   Zachariasz R., Bochenek D., Archives of Metallurgy and Materials. 54, 895 (2009).

[56]   Zhang S.J., Xia R., Shrout T.R., Zang G.Z., Wang J.F., J. Appl. Phys. 100, (104108-1-6) (2006).

[57]   Zhang S., Dong X., Kojima S., Jpn. J. Appl. Phys. 36, 2994 (1997). https://doi.org/10.1143/JJAP.36.2994

[58]   Zhao Z., Buscaglia V., Viviani M., Buscaglia M.T., Mitoseriu L., Testino A., Nygren M., Johnsson M., Nanni P., Phys. Rev. B. 70(2), 8 (2004).

Materials Research Forum LLC
doi: http://dx.doi.org/10.21741/9781945291951

Chapter 2

Synthesis and experimental techniques

Abstract

This chapter presents sample preparation methods for five piezoceramics. It provides and discusses the working principles of the characterization techniques such as, powder X-ray diffraction, scanning electron microscopy, energy dispersive X-ray spectroscopy, UV-visible spectroscopy, dielectric, ferroelectric and piezoelectric measurements.

Keywords: Solid State Reaction, Calcination, Ball Milling, Sintering, Pelletizing, Poling, Dipoles

Contents

Materials Research Forum LLC
doi: http://dx.doi.org/10.21741/9781945291951

2.   Synthesis and experimental techniques

Synthesis of lead-free ceramics is most important in industrial applications and ceramics related areas of materials science. In the work reported in this book, the lead-free piezoceramics have been synthesized using the solid-state reaction technique. This technique involves ball milling, calcination, shaping and sintering. A general discussion on solid-state reaction technique is given in this chapter.

This chapter also deals with different techniques for characterising of powders with X-ray diffraction, scanning electron microscopy, elemental dispersive X-ray spectroscopy, UV-visible spectroscopy and the techniques used for the determination of the dielectric, ferroelectric and piezoelectric properties of the ceramics.

2.1 Synthesis of ceramics

2.1.1 Solid state reaction (SSR)

A solid solution is prepared by mixing two or more compounds of starting chemicals. This method is most widely used for the synthesis of polycrystalline bulk ceramics. Solid-state reaction method provides large range of selection of starting materials like oxides, carbonates, *etc.* Solids usually do not react with each other at room temperature. Hence, they are heated to high temperatures for the proper reaction to take place at an appreciable rate. The solid reactants can react chemically without the presence of any solvent at high temperatures yielding a product which is stable. The advantage of SSR method is the ability to produce structurally pure materials with desired properties depending on the final sintering temperatures. In the solid-state reaction method, the starting chemicals are weighed according to the stoichiometry of the compound with due consideration for impurity and moisture contents. The basic steps needed for the solid-state reaction technique are mixing, ball milling, calcination, pelletizing and final sintering (Narayan *et al.*, 2009; Moulson and Herbert, 2003). The flow chart describes the various stages of processing of samples by the solid-state reaction method which includes sintering, grinding and electroding as shown in figure 2.1.

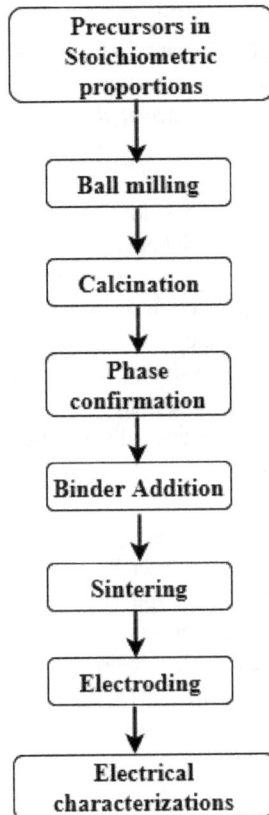

Figure 2.1 Steps involved in solid-state reaction technique of preparing ceramic samples.

Materials Research Forum LLC
doi: http://dx.doi.org/10.21741/9781945291951

2.1.1.1 Mixing and ball milling

For the preparation of a compound from raw materials, particles of the raw materials need to react and diffuse into one another. Hence, all constituents need to be mixed homogenously. Ball milling is a process in which the size of larger particles is reduced to smaller particles by mechanical force. Ball milling is the most common way to achieve the size reduction and involves the operations such as grinding, crushing and milling. Laboratory ball mills, usually referred as ball mills consist of a rotating horizontal cylinder that is partly filled with balls and the particles to be ground. The speed of rotation of the mill is an important variable since it influences the trajectory of the balls and the mechanical energy supplied to the powder. The difference in speeds between the balls and grinding jars produces an interaction between frictional and impact forces, which releases high dynamic energies (Jafee *et al.*, 1971).

2.1.1.2 Calcination

Calcination process involves chemical reactions between the starting precursor chemicals used. During the process, either the partial or complete phase of the compound is formed. Calcination is the process in which a material is heated to a temperature below its melting point to effect the thermal decomposition or the phase transition other than melting point or removal of a volatile fraction. Calcination process also helps in homogenizing the materials and reducing the shrinkage during the subsequent sintering process of the finally shaped samples. Calcined powders are usually coarser and harder than mixed milled powders and the particle size generally increases with increase in calcination temperature (Narayan *et al.*, 2009).

2.1.1.3 Pelletizing

Pelletizing is used for compaction of powder involving uniaxial pressure applied to the powder placed in a die between two rigid punches. The calcined powder is fed into the die cavity. Then, the upper punch moves down and presses the powder with a predetermined pressure developed using a hydraulic press. During pelletizing process, lateral cracks may be developed in the pellets. To overcome this problem, an organic binder (PVA) is incorporated into the calcined powder, for giving sufficient strength to dry shapes.

2.1.1.4 Sintering

Sintering is defined as a process by which a powder compact is transformed to a strong, dense ceramic body upon heating (Matsuo and Sasaki, 2009). Sintering of ceramic materials is the consolidation of ceramic powders by heating the pellets to a high

Materials Research Forum LLC
doi: http://dx.doi.org/10.21741/9781945291951

temperature below the melting point. When thermal energy is applied to the pellets, the pellets are densified and the average grain size is increased, this process is called densification and grain growth. Highly dense ceramics are required for most electrical applications. Thus, two important criteria to optimize sintering temperature are phase purity and sample density. Generally, the pellets sintered in a closed alumina crucible to reduce the loss of volatile substances (Buckner, 1972; Chinor, 2000).

2.1.1.5 Electrode deposition

The silver contact is mainly used in the measurements of electrical properties like ferroelectric and piezoelectric properties. The sintered pellet is polished to a smooth finish and cleaned with alcohol. Then, silver paste is applied on both sides of the sample. Silver electroding on the samples gives the metal-insulator-metal structure to the pellets. The pellets act as a dielectric medium between the two parallel metallic plates.

2.1.1.6 Electrical poling

Poling is a process during which a high electric field is applied on the piezoelectric samples to force the electric domains to reorient in the direction of the applied electric field. Before poling, the ferroelectric ceramics will not possess any piezoelectric properties owing to the random orientation of the ferroelectric domains in the ceramics. An external electric field is applied to the ferroelectric material at a temperature less than the Curie temperature (T_C) for a certain length of time, which gives better domain rearrangement in the ferroelectric materials. When the poling temperature is too high, problems may arise as the increase in electrical conductivity and the consequent increase in leakage current would result in sample breakdown during the period of poling. After poling, the electric field is removed and a remnant polarization is maintained within the material and finally the poled pellet exhibits piezoelectric and ferroelectric effects (Sessler, 1987).

In the work reported in this book, the weight measurements were done using a digital balance (Model MK200E) with a readability of 0.001 gm. The starting materials were initially mixed using laboratory ball mill (figure 2.2(a)) which consists of an agate jar (250 ml) and agate balls ranging from 1 cm and 0.6 cm diameter were utilized for grinding and mixing the raw chemicals effectively. The ground powder samples were made as disc shaped pellets using a hydraulic press which is able to provide the maximum pressure of 10 tons (figure 2.2(b)). Then, the pellets were carefully transferred to boat type crucible made up of alumina, which withstands a temperature of 1650°C.

(a)

(b)

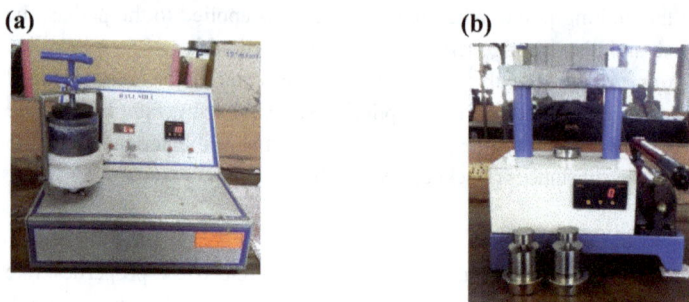

Figure 2.2 Photographs of experimental tools *(a)* Laboratory ball mill and *(b)* Pelletizer.

Figure 2.3 shows the high temperature tubular furnace equipped with Nippon/Eurotherm PID programmable controller. This furnace has the working temperature limit up to 1600°C with one degree accuracy of dwell temperature and has a rapid heating and cooling rate of 5°C/min. In the work reported in this book, calcinations and sintering processes were effectively carried out using the above mentioned tubular furnace.

To characterize the dielectric, ferroelectric and piezoelectric properties of the pellets, both faces of the sintered pellets were coated with silver paste. For the piezoelectric measurements, the pellets were first poled in silicone oil at 40-100°C under a d.c electric field in the range ~3-4 kV/mm for 30 min. Poling is a process during which a high electric field is applied by high voltage d.c power supply (H5K02N, Aplab instruments) to the piezoelectric samples to force the domains to reorient in the direction of the applied electric field. The poling setup is shown in figure 2.4.

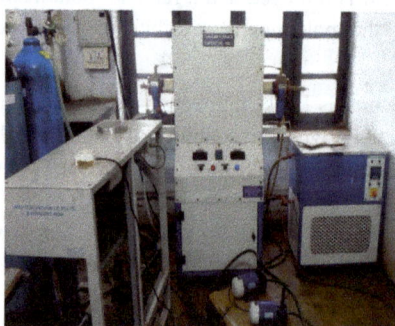

Figure 2.3 High temperature tubular finance for sintering purpose
(temperature range up to 1600°C).

Figure 2.4 *Photograph of the poling setup.*

2.2 Preparation of lead-free solid solutions

2.2.1 Introduction

Piezoelectric materials form an important class of functional materials that can transduce mechanical energy to electrical energy and vice versa. Lead-based piezoelectric perovskite materials are well known for their excellent piezoelectric properties, which are extensively used in industrial applications. The most widely used piezoelectric ceramic is lead zirconate titanate ($Pb(Zr_{1-x}Ti_x)O_3$). One critical disadvantage of PZT is that it contains more than 60 wt. % of lead content, which pollutes the environment strongly and also the health of humans. The large content of Pb creates hazards during preparations and is potentially environmentally toxic during disposal. Currently, there is high interest in the development of lead-free materials due to health and environmental concerns. During the last several years, lead-free piezoelectric materials have been developed to replace the lead-based materials.

In the work reported in this book, lead-free ceramics have been prepared by conventional solid-state reaction method. The synthesis of lead-free ceramics requires special handling of the starting ceramic powders due to volatility of alkaline elements such as sodium, potassium and bismuth based oxides powders. The raw compounds easily decompose under high sintering temperature. When the sintering temperature is above 1000°C, the evaporation of K_2O, Bi_2O_3 and Na_2O degrades the resistivity and the piezoelectric properties. However, it is very difficult to control the evaporation of K_2O, Bi_2O_3 and Na_2O by muffling. Therefore, elaboration procedures are needed to produce high quality lead-free ceramics in a reproducible way. Then, suitable dopants to be substituted in (Na,

Materials Research Forum LLC
doi: http://dx.doi.org/10.21741/9781945291951

$K)NbO_3$-based ceramics, such as $BaTiO_3$ and Sb^{5+}. The following five series have been chosen for the work reported in this book;

(i) $(1-x)(Na_{1-y}K_yNbO_3)$-$xBaTiO_3$ (x=0.1, 0.2, y=0.01, 0.05) (NKN-BT)

(ii) $(1-x)(Na_{1-y}K_yNb_{1-z}Sb_zO_3)$-$xBaTiO_3$ (x=0.1, 0.2; y=0.03, 0.05; z=0.05, 0.1) (NKNS-BT)

(iii) $(1-x)(Na_{0.5}Bi_{0.5}TiO_3)$-$xBaTiO_3$ (x=0.00, 0.04, 0.08, 0.12) (NBT-BT)

(iv) $(1-x)(K_{0.5}Bi_{0.5}TiO_3)$-$xBaTiO_3$ (x=0.00, 0.08, 0.12) (KBT-BT)

(v) $(1-x)Ba(Zr_{0.2}Ti_{0.8})O_3$-$x(Ba_{0.7}Ca_{0.3})TiO_3$, (x=0.4, 0.5, 0.6) (BZT-BCT)

In the work reported in this book, NKN-BT, NKNS-BT, NBT-BT, KBT-BT and BZT-BCT solid solutions with various compositional ranges were synthesized by conventional solid-state reaction route. High purity ($\geq$ 99.99%, Alfa aeser) powders were used according to the stoichiometry. Table 2.1 presents the molecular weights and purity of the starting chemicals.

Table 2.1 Molecular weights of chemicals used.

Chemicals	Molecular weight (gm/mol)	Purity (%)
Na_2CO_3	105.988	99.997
K_2CO_3	138.205	99.997
Nb_2O_5	265.810	99.9985
Sb_2O_3	291.520	99.999
Bi_2O_3	465.96	99.99
$BaCO_3$	197.340	99.99
TiO_2	79.866	99.99
ZrO_2	123.218	99.99
$CaCO_3$	100.086	99.95

2.2.2   Preparation of $(1-x)(Na_{1-y}K_y)NbO_3$-$xBaTiO_3$.

The lead-free $(1-x)(Na_{1-y}K_y)NbO_3$-$xBaTiO_3$ ceramics with the compositions x=0.1, 0.2; y=0.01, 0.05 were prepared by conventional solid-state reaction method. The starting raw compounds, Na_2CO_3, K_2CO_3, Nb_2O_5, $BaCO_3$ and TiO_2 powders were weighed according to the stoichiometric ratio. The chemical equation of the SSR is as follows;

$$(1-x)\left\{\frac{(0.5)}{2}(1-y)Na_2CO_3 + \frac{(0.5)}{2}\,y\,K_2O_3 + Nb_2O_5\right\} + xBaCO_3 + xTiO_2 \rightarrow$$

$$(1-x)(Na_{0.5}Bi_{0.5}TiO_3) - xBaTiO_3 + CO_2 \uparrow \tag{2.1}$$

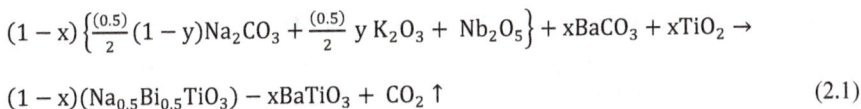

Stoichiometric quantities of precursor materials were taken according to equation (2.1). Table 2.2 gives the actual quantities of raw powders that were used for the synthesis of $(1-x)(Na_{1-y}K_y)NbO_3-xBaTiO_3$, (x=0.1, 0.2; y=0.01, 0.05). All powder samples were mixed and ball milled with agate balls for 2 h using laboratory ball mill. The ball milled power samples were calcined at 950°C for 4 h in air atmosphere using programmable tubular furnace. Again, the calcined powders were milled with an agate ball mill with agate balls for about 2 h. The calcined powders were mixed thoroughly with 2 wt% polyvinyl alcohol (PVA) binder solution. The powders were then pressed into discs of 12 mm diameter and 1 mm thickness using a hydraulic press. These pellets were then finally sintered at 1250°C for 2 h in air with a heating rate of 5°/min. The sintered $(1-x)(Na_{1-y}K_y)NbO_3-xBaTiO_3$, (x=0.1, 0.2; y=0.01, 0.05) pellets are shown in figure 2.5.

Table 2.2 *Quantities of starting materials used to synthesize $(1-x)(Na_{1-y}K_y)NbO_3-xBaTiO_3$, (x=0.1, 0.2; y=0.01, 0.05).*

Concentration (x, y)	Na$_2$CO$_3$ (gm)	K$_2$CO$_3$ (gm)	Nb$_2$O$_5$ (gm)	BaCO$_3$ (gm)	TiO$_2$ (gm)
x=0.1; y=0.01	0.9448	0.0124	2.1530	0.3946	0.1597
x=0.1; y=0.05	0.9061	0.0621	2.1530	0.3946	0.1597
x=0.2; y=0.01	0.8394	0.0110	1.9138	0.7893	0.3190
x=0.2; y=0.05	0.8055	0.0552	1.9138	0.7893	0.3190

Figure 2.5 *The synthesized samples of $(1-x)(Na_{1-y}K_y)NbO_3-xBaTiO_3$, (x=0.1, 0.2; y=0.01, 0.05).*

Materials Research Forum LLC
doi: http://dx.doi.org/10.21741/9781945291951

2.2.3 Preparation of $(1-x)(Na_{1-y}K_y)(Nb_{1-z}Sb_z)O_3$-xBaTiO$_3$

Lead-free $(1-x)(Na_{1-y}K_y)(Nb_{1-z}Sb_z)O_3$-xBaTiO$_3$, (x=0.1, 0.2; y=0.03, 0.05; z=0.05, 0.1) ceramics were prepared by solid-state reaction method. The high purity chemicals including, Na_2CO_3, K_2CO_3, Nb_2O_5, Sb_2O_3 and BaTiO$_3$ were used as raw materials. The appropriate weights of the raw powders are given in Table 2.3. The chemical equation of the SSR is as follows;

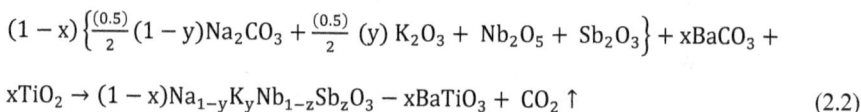

$$(1-x)\left\{\frac{(0.5)}{2}(1-y)Na_2CO_3 + \frac{(0.5)}{2}(y)K_2O_3 + Nb_2O_5 + Sb_2O_3\right\} + xBaCO_3 +$$

$$xTiO_2 \rightarrow (1-x)Na_{1-y}K_yNb_{1-z}Sb_zO_3 - xBaTiO_3 + CO_2 \uparrow \qquad (2.2)$$

Stoichiometric weights of all the powders were mixed and ball milled for 3 h using agate balls. The mixed $(1-x)(Na_{1-y}K_y)(Nb_{1-z}Sb_z)O_3$-xBaTiO$_3$ powders were calcined at 1000°C for 3 h. The calcined powders were made as pellets by adding 2 wt. % polyvinyl alcohol (PVA) as a binder, using a hydraulic press to obtain discs with 1 mm thickness and 12 mm diameter. Finally, the pellets were sintered at 1250°C for 2 h with a heating rate of 5°C/min. The synthesized $(1-x)(Na_{1-y}K_y)(Nb_{1-z}Sb_z)O_3$-xBaTiO$_3$, (x=0.1, 0.2; y=0.03, 0.05; z=0.05, 0.1) pellets are shown in figure 2.6.

Table 2.3 *Quantities of starting materials used to synthesize*
$(1-x)(Na_{1-y}K_y)(Nb_{1-z}Sb_z)O_3$-xBaTiO$_3$, (x=0.1, 0.2; y=0.03, 0.05; z=0.05, 0.1)

Concentration (x, y, z)	Na$_2$CO$_3$ (gm)	K$_2$CO$_3$ (gm)	Nb$_2$O$_5$ (gm)	Sb$_2$O$_3$ (gm)	BaCO$_3$ (gm)	TiO$_2$ (gm)
x=0.1, y=0.03, z=0.05	0.9252	0.0373	2.2720	0.1311	0.3946	0.1597
x=0.1, y=0.05, z=0.1	0.9061	0.0621	0.2392	0.2623	0.3946	0.1597
x=0.2, y=0.03, z=0.05	0.8224	0.0331	2.0201	0.1166	0.7893	0.3194
x=0.2, y=0.05, z=0.1	0.8055	0.0552	0.2126	0.2332	0.7893	0.3194

| x=0.1, y=0.03, z=0.05 | x=0.1, y=0.05, z=0.1 | x=0.2, y=0.03, z=0.05 | x=0.2, y=0.05, z=0.1 |

Figure 2.6 The synthesized samples of $(1-x)(Na_{1-y}K_y)(Nb_{1-z}Sb_z)O_3-xBaTiO_3$, $(x=0.1, 0.2;$
$y=0.03, 0.05; z=0.05, 0.1)$.

2.2.4 Preparation of $(1-x)(Na_{0.5}Bi_{0.5})TiO_3-xBaTiO_3$

The lead-free ceramics $(1-x)(Na_{0.5}Bi_{0.5})TiO_3-xBaTiO_3$, $(x=0.00, 0.04, 0.08, 0.12)$ were prepared by solid-state reaction method. Na_2CO_3, Bi_2O_3, TiO_2 and $BaCO_3$ were used as starting chemicals to prepare the solid solution. The chemical equation of the SSR is as follows;

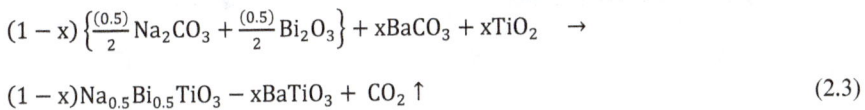

$$(1-x)\left\{\frac{(0.5)}{2}Na_2CO_3 + \frac{(0.5)}{2}Bi_2O_3\right\} + xBaCO_3 + xTiO_2 \rightarrow$$

$$(1-x)Na_{0.5}Bi_{0.5}TiO_3 - xBaTiO_3 + CO_2 \uparrow \qquad (2.3)$$

The individual weights of powders are given in Table 2.4. The raw powders were weighed in the stoichiometric ratio of $(1-x)(Na_{0.5}Bi_{0.5})TiO_3-xBaTiO_3$. The weighed powders were ball-milled with agate balls for 2 h. The mixed powders were calcined at 850°C for 2 h. After a second milling, powders of all the compositions were further mixed with 2 wt. % of polyvinyl alcohol (PVA) as binder and the samples were pressed using hydraulic press with a die of 12 mm diameter and 1 mm thickness. Finally, these pellets were sintered at 1150°C for 2 h in a programmable furnace. The synthesized pellets of $(1-x)(Na_{0.5}Bi_{0.5})TiO_3-xBaTiO_3$, $(x=0.00, 0.04, 0.08, 0.12)$ are shown in figure 2.7.

Table 2.4 *Quantities of starting materials used to synthesize*
$(1-x)(Na_{0.5}Bi_{0.5})TiO_3$-$xBaTiO_3$, (x=0.00, 0.04, 0.08, 0.12).

Concentration (x)	Na₂CO₃ (gm)	Bi₂O₃ (gm)	BaCO₃ (gm)	TiO₂ (gm)
x=0.00	0.5299	2.3298	-	1.5973
x=0.04	0.5087	2.2366	0.1578	1.5972
x=0.08	0.4875	2.1430	0.3157	1.5967
x=0.12	0.4663	2.0502	0.4736	1.5969

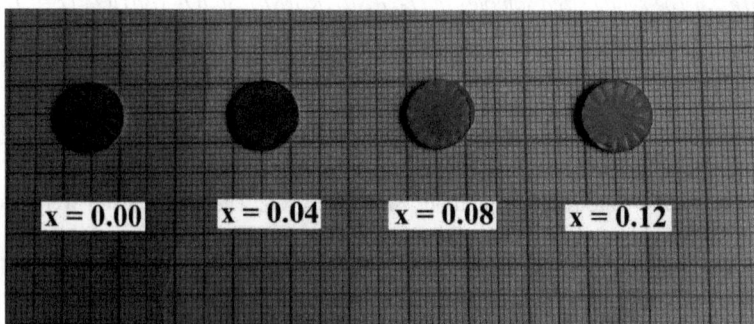

Figure 2.7 *The synthesized samples of $(1-x)(Na_{0.5}Bi_{0.5})TiO_3$-$xBaTiO_3$,*
(x=0.00, 0.04, 0.08, 0.12).

2.2.5 Preparation of $(1-x)(K_{0.5}Bi_{0.5})TiO_3$-$xBaTiO_3$

Potassium bismuth titanate-barium titanate ceramics were prepared using the high purity raw compounds of K_2CO_3, Bi_2O_3, TiO_2 and $BaTiO_3$. The individual weights of stoichiometric molar ratio powders were given in Table 2.5. The chemical equation of the SSR is as follows;

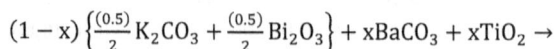

$$(1 - x)\left\{\frac{(0.5)}{2}K_2CO_3 + \frac{(0.5)}{2}Bi_2O_3\right\} + xBaCO_3 + xTiO_2 \rightarrow$$

$$(1 - x)K_{0.5}Bi_{0.5}TiO_3 - xBaTiO_3 + CO_2 \uparrow \tag{2.4}$$

Lead-free Piezo-Ceramic Solid Solutions, R. Saravanan   Materials Research Forum LLC
Materials Research Foundations **41** (2018)   doi: http://dx.doi.org/10.21741/9781945291951

The raw compounds were ball milled for 2 h with agate balls and then the powder samples were calcined at 800°C for 2 h. The $(1-x)(K_{0.5}Bi_{0.5})TiO_3\text{-}xBaTiO_3$ powders were pressed uniaxially at 200 MPa with 2 wt.% polyvinyl alcohol added as a binder. Finally, the $(1-x)(K_{0.5}Bi_{0.5})TiO_3\text{-}xBaTiO_3$ phase was obtained when the sintering process was performed at 1050°C for 3 h. The synthesized $(1-x)(K_{0.5}Bi_{0.5})TiO_3\text{-}xBaTiO_3$, (x=0.00, 0.08, 0.12) pellets are shown in figure 2.8.

Table 2.5 *Quantities of starting materials used to synthesize*
$(1-x)(K_{0.5}Bi_{0.5})TiO_3\text{-}xBaTiO_3$, (x=0.00, 0.08, 0.12)

Concentration (x)	K_2CO_3 (gm)	Bi_2O_3 (gm)	$BaCO_3$ (gm)	TiO_2 (gm)
x=0.00	1.3820	4.6596	-	1.5973
x=0.08	0.6357	2.1434	0.3157	1.5972
x=0.12	0.6081	2.0502	1.4056	0.1916

Figure 2.8 *The synthesized samples of $(1-x)(K_{0.5}Bi_{0.5})TiO_3\text{-}xBaTiO_3$,*
(x=0.00, 0.08, 0.12).

2.2.6 Preparation of $(1-x)Ba(Zr_{0.2}Ti_{0.8})O_3\text{-}x(Ba_{0.7}Ca_{0.3})TiO_3$

Polycrystalline ceramics of $(1-x)Ba(Zr_{0.2}Ti_{0.8})O_3\text{-}x(Ba_{0.7}Ca_{0.3})TiO_3$, (x=0.4, 0.5, 0.6) were synthesized by conventional solid-state reaction method. The chemical equation of the SSR is as follows;

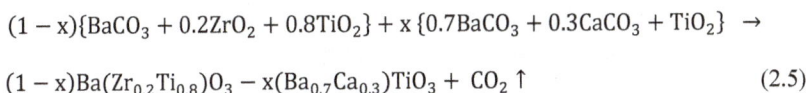

$$(1-x)\{BaCO_3 + 0.2ZrO_2 + 0.8TiO_2\} + x\{0.7BaCO_3 + 0.3CaCO_3 + TiO_2\} \rightarrow$$

$$(1-x)Ba(Zr_{0.2}Ti_{0.8})O_3 - x(Ba_{0.7}Ca_{0.3})TiO_3 + CO_2 \uparrow \qquad (2.5)$$

The individual weights of powders are given in Table 2.6. Commercially available powders of $BaCO_3$, $CaCO_3$, TiO_2 and ZrO_2 were used as starting materials. The raw compounds were mixed with stoichiometric proportions and ball-milled for 12 h. Subsequently, the mixed powders were calcined in alumina crucibles at 1350°C for 2 h and ball-milled again for 8 h. The calcined powders were pelletized using a hydraulic press with die of 10 mm diameter and 1 mm thickness. Finally, the pellets were sintered at 1450°C in air for 3 h. The sintered $(1-x)Ba(Zr_{0.2}Ti_{0.8})O_3$-$x(Ba_{0.7}Ca_{0.3})TiO_3$, (x=0.4, 0.5, 0.6) pellets are shown in figure 2.9.

Table 2.6 *Quantities of starting materials used to synthesize*
$(1-x)Ba(Zr_{0.2}Ti_{0.8})O_3$-$x(Ba_{0.7}Ca_{0.3})TiO_3$, (x=0.4, 0.5, 0.6).

Concentration (x)	BaCO₃ (gm)	ZrO₂ (gm)	CaCO₃ (gm)	TiO₂ (gm)
x=0.4	2.8943	0.2464	0.2001	1.1713
x=0.5	2.7956	0.2053	1.5511	1.1979
x=0.6	2.6969	1.642	0.3002	1.2245

Figure 2.9 *The synthesized samples of $(1-x)Ba(Zr_{0.2}Ti_{0.8})O_3$-$x(Ba_{0.7}Ca_{0.3})TiO_3$ (x=0.4, 0.5, 0.6).*

2.3   Characterization methods and instrumentations

This section describes the characterization techniques and working principles of various instruments used to characterize the lead-free ceramic solid solutions. The characterization techniques used in the work reported in this book mainly include;

(i)   X-ray diffraction (XRD)

(ii)   Scanning electron microscopy (SEM)

(iii)   Energy dispersive X-ray spectroscopy (EDS)

Lead-free Piezo-Ceramic Solid Solutions, R. Saravanan   Materials Research Forum LLC
Materials Research Foundations **41** (2018)   doi: http://dx.doi.org/10.21741/9781945291951

(iv)   UV-visible spectroscopy (UV-vis)

(v)   Dielectric measurements

(vi)   P-E loop measurements (polarization versus electric field loop)

(vii)   Piezoelectric constant measurements

2.3.1 X-ray diffraction (XRD)

X-ray diffraction (XRD) is a technique to determine the structural properties of crystalline samples. This technique is based on the constructive interference of monochromatic X-rays from a crystalline sample. The X-ray diffraction technique is used to confirm the phase purity and analyze the crystal structure of the prepared samples. The basic components of the X-ray diffractometer are X-ray tube (X-ray generator), goniometer and detector (Azaroff, 1968). A beam of accelerated electrons produced as a result of thermionic emission from a tungsten filament is allowed to fall on a metal target usually copper (Cu). These accelerated electrons knock off the electron from the target atom (1s in the case of Cu). X-ray photon is released when an electron from higher orbits jump to the vacant inner orbit. The energy of the photon ejected out is characteristic of the target material. In the case of copper, the characteristic wavelength is 1.5406 Å ($CuK\alpha_1$) or 1.5443 Å ($CuK\alpha_2$). These two different wavelengths depend upon the spin state of the 2p orbital. X-rays are electromagnetic radiation with typical photon energies in the range of 100 eV-100 keV. For diffraction applications, only short wavelength X-rays in the range of a few angstroms to 0.1 angstroms are used, because the wavelength of X-rays is comparable to the size of atoms and hence is ideally suited for probing the structural arrangement of atoms and molecules in a wide range of materials. The X-rays generated by a cathode ray tube are filtered to produce monochromatic radiation, collimated and directed towards the sample.

Apart from the characteristic wavelength, a continuous radiation known as Bremsstrahlung radiation is also produced and a Ni filter in monochromator is used to remove the Bremsstrahlung radiation. When a material is exposed to an incident X-ray beam, the X-rays interact with the atoms arranged in a periodic fashion, act as secondary sources and reradiate the X-rays in all directions. In the periodic array, constructive and destructive interference described by W.L. Bragg, may result in the reflection from sets of parallel planes of lattice points. If a beam of X-ray interacts with two successive crystallographic planes of given Miller indices, the path difference (2d) between two reflected beams is equal to $2d_{hkl} \sin \theta$ shown in figure 2.10. The interference is constructive when the phase shift is a multiple of 2π; this condition can be expressed by Bragg's law (Bragg, 1913):

Materials Research Forum LLC
doi: http://dx.doi.org/10.21741/9781945291951

$$n\lambda = 2d \sin\theta \tag{2.6}$$

where n is an integer determined by the order given, λ is the wavelength of the X-ray, d is the spacing between the planes in the atomic lattice and θ is the angle between the incident ray and the scattering planes.

2.3.1.1 Instrumentation

An X-ray diffractometer consists of source of radiation, monochromator to choose the wavelength, slits to adjust the shape of the beam, sample and detector (Cullity and Stock, 2001). A goniometer is used for fine adjustment of the sample and the detector position. X-ray contains several components such as K_α and K_β. The specific wavelengths are characteristic of the target material (Cu, Fe, Mo, Cr). Monochromators and filters are used to absorb the unwanted emission with wavelength K_β, while allowing the desired wavelength, K_α to pass through. The filtered X-rays are collimated and directed into the sample as shown in the figure 2.11. When the incident X-ray beam strikes on powder sample, the diffraction occurs in every possible orientation of 2θ. The diffracted X-ray beam is detected by movable detector, which is connected to a recorder. The counter is set to scan over a range of 2θ values at a constant angular velocity. The 2θ range of 5 to 120 degrees is sufficient to cover the most useful part of the powder pattern. A detector records the diffracted X-ray signal and converts the signal to a count rate which is then fed to a computer monitor. The sample must be ground to fine powder before loading it in the glass sample holder.

In the work reported in this book, the powder X-ray diffraction measurements were carried out at room temperature using a Bruker AXS D8 Advance (Karlsruhe, Germany) instrument with CuK_α radiation (λ=1.54056 Å) at Sophisticated Analytical Instrument Facility (SAIF), STIC, Cochin University, Cochin, India.

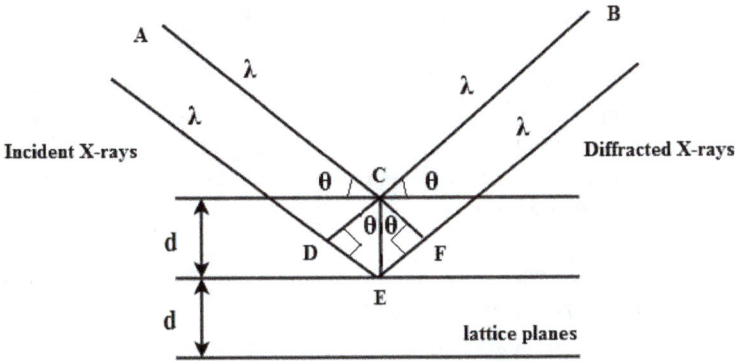

Figure 2.10 *Diffraction of X-ray by lattice planes (Bragg's law).*

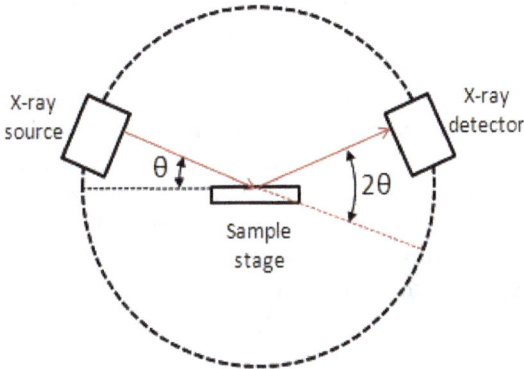

Figure 2.11 *Schematic diagram of X-ray diffractometer.*

2.3.2 Scanning electron microscopy (SEM)

Scanning electron microscopy is an extremely useful tool to analyze the surface morphology of the samples which offers a better resolution than the optical microscope. In SEM, a sample is scanned by a highly energetic electron beam. The electrons are produced in an electron gun through thermionic emission and accelerated in vacuum by the application of high electric filed. When the electron beam impinges on the specimen, several types (Back scattered electrons, secondary electrons, auger electrons, X-rays, cathodoluminiscence, etc.) of signals are generated in the instrument. The two signals used to generate SEM images are secondary electrons (SE) and backscattered electrons (BSE). The interaction of incident electrons with the sample results in the emission of secondary electrons from the atoms of the sample. Most of the electrons are scattered at large angles (from 0° to 180°) when they interact with the positively charged nucleus. These elastically scattered electrons are usually called backscattered electrons (BSE). The secondary, backscattered electron signals are detected by a detector and a magnified image is produced and displayed on the screen (Lawes, 1987).

In a SEM instrument, the electron gun produces a stream of monochromatic electrons. Figure 2.12 shows schematic diagram of the phenomenon. The objective lens is used to focus the beam on to the sample. The raster scanning of the sample is achieved by changing the current in the scanning coil. At each scanning point, secondary and back-scattered electrons are generated and these electrons are detected by the detector. The counted electrons in the detector are used for determining the relative intensity of the pixel point in the final processed image. For the higher atomic number (Z) materials, back-scattered electrons are higher, giving rise to a higher intensity in the image. The image formed by secondary and back-scattered electrons contains light and dark shades. The X-rays emitted from the transition of electron from higher energy level to the vacant inner shell also provide the information of the elements of the parent compound. A quantitative and qualitative elemental analysis can be performed using SEM images. However, this information is limited to the small area of the sample exposed to the incident electron beam. Therefore, different regions have to be analysed in order to extract the information over the whole sample area.

In the work reported in this book, the SEM measurements were done using JEOL Model JSM-6390LV and Carl Zeiss EVO 18 microscopes at Sophisticated Analytical Instrument Facility (SAIF), STIC, Cochin University, Cochin, India and International research centre (IRC), Kalasalingam University, Krishnankoil, Tamil Nadu, India, respectively.

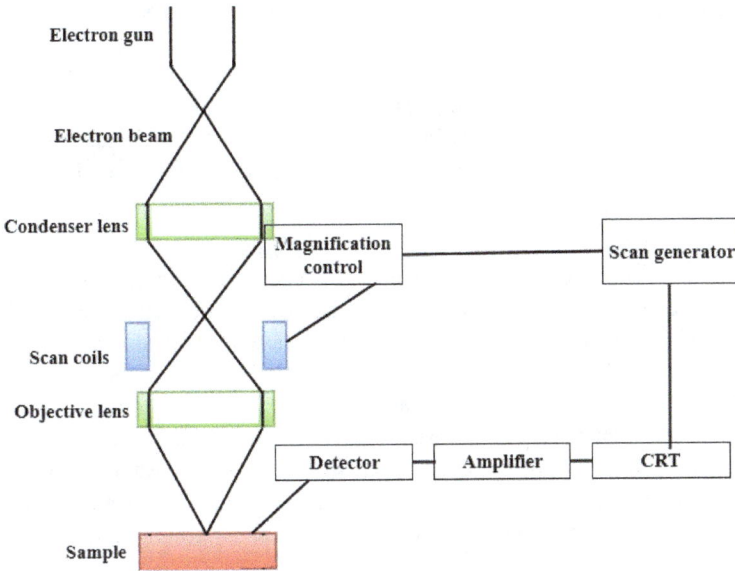

Figure 2.12 Schematic diagram of scanning electron microscope (SEM).

2.3.3 Energy dispersive X-ray spectroscopy (EDS)

Energy dispersive X-ray spectroscopy (EDS) is a technique used for quantitative elemental analysis. The principle of EDS is based on the atomic structure and energy levels of the elements. High-energy electrons can interact with atoms in a sample in many ways, including stimulating the emission of X-rays. In this case, when an electron strikes an atom, it ejects an electron originally positioned in an inner shell (K shell). To return the atom to its lowest energy state, this vacancy is immediately filled by an electron moving from a higher-energy shell in the atom. The high energy electron must release some of its energy in the form of X-rays. As a consequence, the energy released (expressed in eV) is equal to the energy difference between the two levels (Figure 2.13). To complicate the situation further, several electrons at several energy levels can potentially occupy the vacancy, with each releasing different amounts of energy. As a result, even a pure sample will emit X-rays at different energies. For example, if an L-shell electron drops, it emits K_α radiation; whereas an M-shell electron emits K_β (the energy difference between M and K is higher than between L and K). However, because the most probable transition is the L to K movement, K_α radiation will always be more intense than K_β radiation. X-rays in the range of 100 eV to 20 keV are readily measured with a Si(Li) or silicon drift detector (SDD) and this range can be extended to 100 keV with an HpGe detector. The energy dispersive approach provides the great practical value of EDS, as it enables access to virtually the entire periodic table (except H, He and Li) (Russ, 1984).

Figure 2.14 shows the basic components of energy dispersive X-ray analysis system. The X-ray photon first creates a charge pulse in a semiconductor detector; the charge pulse is then converted into a voltage pulse whose amplitude reflects the energy of the detected X-ray. Finally, this voltage pulse is converted into a digital signal, which causes one count to be added to the corresponding channel of a multichannel analyzer. The accumulated counts from the sample produce spectral peaks. The quantitative analysis of an acquired spectrum comprises at least five steps: (1) accounting for spurious peaks; (2) identification of the elements giving rise to the spectrum; (3) removal of the background; (4) resolution of spectral peaks; and (5) computation of element concentration, a process that involves accounting for interelement effects within the systems sample.

In the work reported in this book, the elemental compositions of the samples were analysed by JEOL Model JED-2300 and Quantax 200 with X-flash-Bruker instrument at Sophisticated Analytical Instrument Facility (SAIF), STIC, Cochin University, Cochin, India and IRC, Kalasalingam University, Krishnankoil, Tamil Nadu, India.

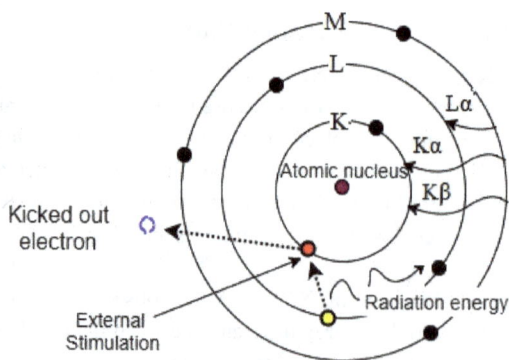

Figure 2.13 The interaction of an electron beam with electrons within an atom.

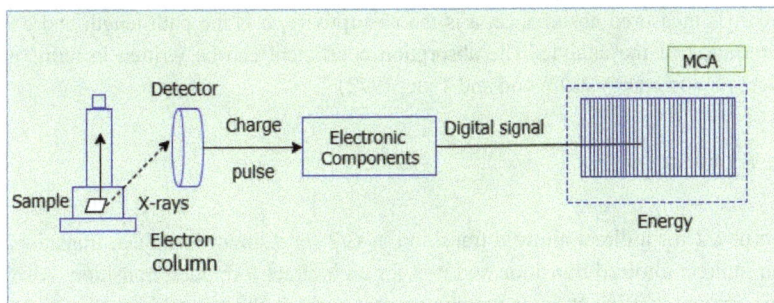

Figure 2.14 Block diagram of Energy dispersive X-ray spectrometer.

2.3.4 UV-visible spectroscopy (UV-vis)

Ultraviolet-visible spectroscopy refers to absorption spectroscopy which ranges in the electromagnetic radiation between 190 nm to 800 nm which is classified into the ultraviolet and visible regions. When sample molecules are exposed to light having an energy (E=hv where E is energy, h is Planck's constant and v is frequency), that matches a possible electronic transition within the molecule, some of the light energies are absorbed as the electron is promoted to a higher energy orbital. The absorption of the material is recorded by the optical spectrometer at each wavelength. The resulting spectrum is presented as a plot of wavelength (λ) versus absorbance (A). The optical properties of materials can be studied with the help of UV-visible spectra.

The position of maximum absorption for a molecule depends on the difference in the energy of the ground state level to that of excited state. Absorption band shows two important characteristic positions of the band which depend on the energy difference between electronic level and intensity which then depend on the interaction between the radiation and electronic system as well as on the energy difference between the ground and excited state. A convenient expression, which relates the absorbance with the path length that the radiation travels within the system and the concentration of the species, can be derived from the Lambert-Beer law and is given in following equation.

$$A = a\,b\,c \qquad\qquad (2.7)$$

where A is measured absorbance, a is the absorptivity, b is the path length and c is the concentration of the analyte. The absorption coefficient can be written in terms of the incident photon energy as (Wood and Tauc, 1972).

$$\alpha h v = K \, (h v - E_g)^n \qquad\qquad (2.8)$$

where n=1/2 for a direct allowed transition, n=3/2 for a direct forbidden transition, n=2 for an indirect allowed transition and n=3 for an indirect forbidden transition. Also, α is the absorption coefficient, hv is the photon energy, E_g is the optical band gap and K is a constant. The band gap is determined by extrapolating the linear portion of the plots of $(\alpha h v)^2$ versus (hv).

2.3.4.1 Instrumentation

The basic components of UV-visible spectrophotometers (Gullapalli and Barron, 2010) are light source, diffraction grating, rotating discs, sample cell and detector as shown in figure 2.15. The radiation light source is the combination of a deuterium lamp for UV region of the spectrum and tungsten or halogen lamp for visible region. The radiation of UV and/or visible light is separated into its component wavelengths by diffraction grating. Then, the monochromatic beam comes from the slit falls onto the rotating disc. The rotating disc is configured with different segments as opaque black, transparent and mirrored segments. If the light hits the transparent segment, it passes through the sample cell and reaches the mirrored segments of the second rotating disc through the reflecting mirror and then collected by the detector. If the light hits the mirrored segments, it passes through reference cell and hits the transparent segment of the second rotating disc and collected by the detector. If the light hits the opaque segment, no light passes into the instrument. The detector converts the observed light into electrical signals and which is read out by the computer. The intensity of the observed pattern mainly depends on the observed electrical signal. The computer calculates absorbance using the observed signals.

In the work reported in this book, UV-visible absorption spectra of the synthesized samples were done using a UV-visible spectrophotometer (Cary 5000, Varian, Germany) in the range of 200-2000 nm wavelengths at Sophisticated Analytical Instrument Facility (SAIF), STIC, Cochin University, Cochin, India.

Lead-free Piezo-Ceramic Solid Solutions, R. Saravanan
Materials Research Foundations **41** (2018)

Materials Research Forum LLC
doi: http://dx.doi.org/10.21741/9781945291951

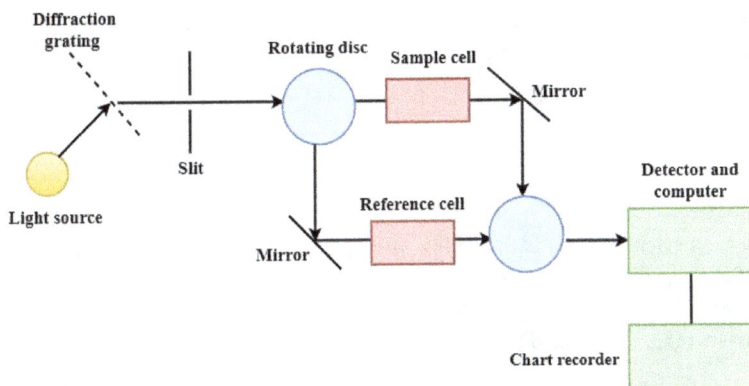

Figure 2.15 Functional Block diagram of a double beam UV-visible spectrophotomete.

2.3.5 Dielectric measurements

The dielectric constant (ε) is defined as the ratio of the permittivity of the material to the permittivity of free space. The dissipation factor is defined as the tangent of the loss angle (tan δ). It is a measure of the amount of electrical energy which is lost through conduction when a voltage is applied across the ceramic sample. The measured parameters are used to understand the behaviour and mechanism of the electric polarizations in the sample. The dielectric properties are measured as a function of temperature at different frequencies using high resolution broad band dielectric spectrometer equipped with a high temperature system. The sample is usually mounted between two electrodes forming a sample parallel-plate capacitor. The schematic diagram of the circuit used in the dielectric constant measurements is shown in figure 2.16. In this circuit, the voltage is applied by a signal generator across the pair of electrodes on the surfaces of the sample capacitor (C_s, with thickness d) and a reference capacitor (C_r). The voltage across the reference capacitor (V_r) and the sample capacitor (V_s) is also measured by two separate voltmeters. Because the two capacitors are connected in series, the charge on the reference capacitor (Q_r) must be the same as the charge over the sample capacitor (Q_s). This means that the charge on the sample can be found by:

$$Q = C_r \times V_r \tag{2.9}$$

Materials Research Forum LLC
doi: http://dx.doi.org/10.21741/9781945291951

The voltage (V_s) is also measured across the sample, allowing for the capacitance of this material to be determined by the following equation:

$$C_s = \frac{Q}{V_s} \tag{2.10}$$

In general, capacitance (C) is the ability of a body to hold the electrical charge, usually stated for a parallel plate capacitor where two parallel metal electrodes, each of one having area A, separated by a distance d and filled by a dielectric materials of permittivity (ε) is represented below,

$$C = \frac{\varepsilon\, A}{d} \tag{2.11}$$

$$C_0 = \frac{\varepsilon_0\, A}{d} \tag{2.12}$$

where, ε and ε_0 represent the permittivity of the material and permittivity of free space respectively. Moreover, the measured capacitance was then changed into dielectric constant using the following formula:

$$\varepsilon_r = \frac{C}{C_0} = \frac{\varepsilon}{\varepsilon_0} \tag{2.13}$$

where, ε_r stands for relative permittivity. Moreover, when a dielectric is placed in the alternating voltage, the electrical energy is absorbed by the material which is dissipated in the form of heat. The dissipation is called as dielectric loss (tan δ). The dielectric loss is related to the imaginary part of the permittivity, however, it is most often described by the tangent of the dielectric loss (tan δ).

$$\tan \delta = \frac{\varepsilon''}{\varepsilon'} \tag{2.14}$$

where δ is the angle between the imaginary component and the real component of the permittivity.

Materials Research Forum LLC
doi: http://dx.doi.org/10.21741/9781945291951

In the work reported in this book, the temperature dependent dielectric properties of the solid solutions were investigated using an impedance analyzer (PSM 1735, N4L) at SSN Engineering College, Chennai, India for different frequencies from room temperature up to 400°C as shown in figure 2.17.

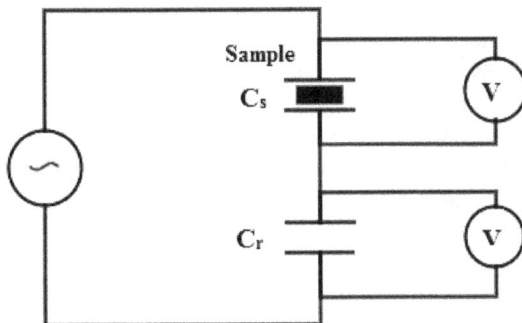

Figure 2.16 *The schematic diagram of the circuit used in the dielectric permittivity measurements.*

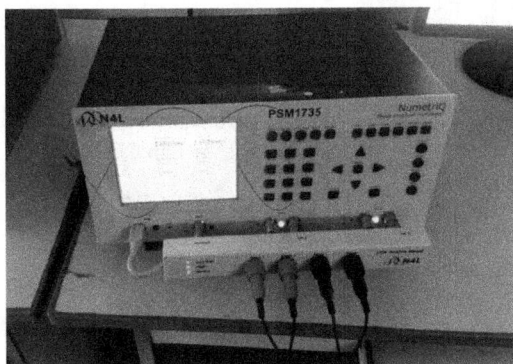

Figure 2.17 *Photograph of dielectric measurement system.*

2.3.6 P-E loop measurements

A ferroelectric material can be analyzed by tracing the polarization versus electric field (P-E) hysteresis loop. The polarization hysteresis measurement based on standard Sawyer-Tower circuit (Sawyer and Tower, 1930) is shown in figure 2.18. In this circuit, a step voltage (V) is applied by signal generator on surface of electrodes of a sample capacitor (C with thickness d). The voltage (V_o) across the standard capacitor (C_o) is measured, and because the two capacitors are connected in series, the charge on the reference capacitor must be same as the charge over the sample capacitor. This means that the charge on the ferroelectric sample can be found by using equation 2.9, as discussed previously. Once the charge on the sample is known, the polarization can be determined through the following equation,

$$P = \frac{Q}{A}$$
(2.15)

where Q is the charge developed on the electrodes of the ferroelectric capacitor and A is the areas of the electrodes. A Sawyer-Tower circuit allows for the determination of a materials polarization as a function of electric field, which rises to a ferroelectric hysteresis loop. In a ferroelectric hysteresis loop measurement, the opposite surfaces of the ceramic samples are coated with a layer of silver paste to make the electrodes. High electric field was applied on disc under silicone oil at room temperature using a computer-controlled function generator and a high voltage amplifier.

In the work reported in this book, the ferroelectric studies were done in a P-E loop tracer, Precision Multiferroic Tester PMF0713-334 (Radiant Technologies, USA) at National Institute of Technology (NIT), Trichy, India as shown in figure 2.19.

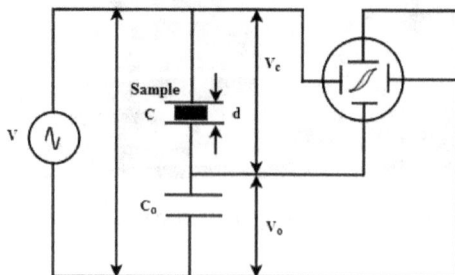

Figure 2.18 *Schematic circuit of the Sawyer-Tower Bridge for measuring the P-E characteristics of ferroelectrics.*

Figure 2.19 Photograph of P-E loop tracer.

2.3.7 Piezoelectric constant (d_{33}) measurements

Piezoelectric constant (d_{33}) is an important parameter to evaluate the piezoelectric properties of a material. The piezoelectric constant is defined by the following equation (Damjanovic, 1998; Haertling, 1999):

$$d_{33} = \left(\frac{D_3}{\sigma_3}\right)_E \tag{2.16}$$

where D_3 is the charge density and σ_3 is the mechanical stress at constant electric field. For the piezoelectric measurements, the sintered pellets are first poled in silicone oil under a d.c electric filed about for 30 min. Figure 2.20 shows a schematic diagram of the piezoelectric measurement system using the piezo-meter. Direct piezoelectric coefficient can be measured by applying fixed force of 0.25 N to the sample using d_{33} meter. The poled pellet sample to be tested is placed between the top and bottom probes. The probe drives the sample vibration and then transfers the signal from the test sample to the measure system.

In the work reported in this book, the piezoelectric constant (d_{33}) of the solid solutions was measured using piezo-d_{33} meter (SINOCERA, YE2730A) at Research Centre and PG Department of Physics, The Madura College, Madurai, Tamil Nadu, India. Figure 2.21 shows the piezo-d_{33} meter instrument used to measure the piezoelectric constant in the work reported in this book.

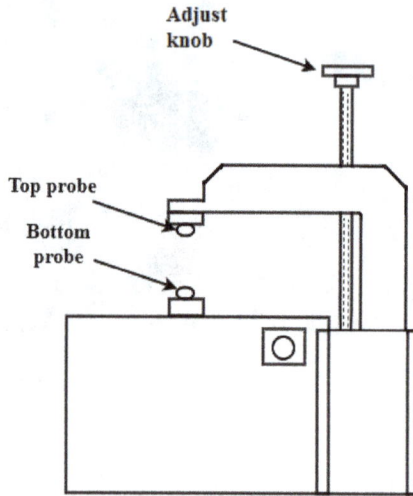

Figure 2.20 *Schematic of piezoelectric constant (d_{33}) measurement.*

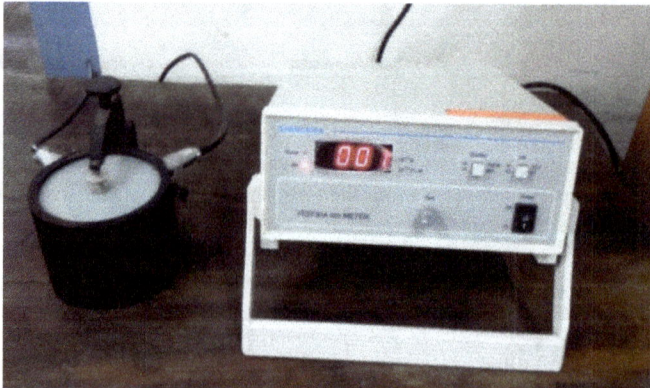

Figure 2.21 *Piezoelectric meter (d_{33}) setup.*

References

[1]   Azaroff L.V., Elements of X-Ray Crystallography, McGraw-Hill, New York, 552, (1968).

[2]   Bragg W.L., Proceedings of the Cambridge Philosophical Society, 17, 43 (1913).

[3]   Buckner D.A., Wilcox P.D., Ceram. Bull. 51, 218-222 (1972).

[4]   Chinor K.U., Ferroelectric Devices, Marcel Dekker INC, New York (2000).

[5]   Cullity B.D., Stock S.R., Elements of X-ray Diffraction, 3rd edition, (Prentice Hall, New Jersy), 167 (2001).

[6]   Damjanovic D., Reports on Progress in Physics. 61, 1267-1324 (1998). https://doi.org/10.1088/0034-4885/61/9/002

[7]   Gullapalli S, Barron A. R, 'Characterization of Group 12-16 (II-VI) Semiconductor Nanoparticles by UV-visible Spectroscopy', OpenStax CNX, June, 2010 Online: Web site. http://cnx.org/content/m34601/1.1/

[8]   Haertling G.H., J. Am. Ceram. Soc. 82, 797-818 (1999). https://doi.org/10.1111/j.1151-2916.1999.tb01840.x

[9]   Jaffe B., Cook W.R., Jeffe H., Piezoelectric ceramics. Academic press, London and New York, Vol. 3 (1971).

[10]   Lawes G., Scanning electron microscopy and X-ray microanalysis: Analytical chemistry by open learning, John Wiley & sons, (1987).

[11]   Moulson A.J., Herbert J.M., Electroceramics: materials, properties and applications, 2nd Edition, Wiley, New York (2003). https://doi.org/10.1002/0470867965

[12]   Narayan H., Alemu H., Macheli L., Rao G., Nanotechnology. 20, 255601 (2009). https://doi.org/10.1088/0957-4484/20/25/255601

[13]   Matsuo Y., Sasaki H., Ceram. Bull. 51, 218 (2009).

[14]   Russ J.C., Fundamentals of Energy Dispersive X-ray Analysis, Butterworths, London (1984).

[15]   Sawyer C.B., Tower C.H., Physical Review. 35, 269-273 (1930). https://doi.org/10.1103/PhysRev.35.269

[16]   Sessler G.M., Electrets, Edited by G M Sessler, 2nd Edition, Springer-Verlag, (1987). https://doi.org/10.1007/3-540-17335-8

[17]   Wood D.L., Tauc J., Phys. Rev. B. 5, 3144 (1972).
https://doi.org/10.1103/PhysRevB.5.3144

Lead-free Piezo-Ceramic Solid Solutions, R. Saravanan   Materials Research Forum LLC
Materials Research Foundations 41 (2018)   doi: http://dx.doi.org/10.21741/9781945291951

Chapter 3

Results

Abstract

Chapter III presents the results obtained from various characterization techniques and analytical methods performed to investigate the five differently doped $BaTiO_3$ lead-free ceramics systems. The plots of experimental X-ray diffraction patterns, Rietveld fitted profiles, SEM micrographs, EDS spectra, Tauc plots, dielectric, ferroelectric measurements, two dimensional, three dimensional and one dimensional electron density line profiles are presented. The tables of structural parameters refined from Rietveld method and optical band gap values with respect to the dopant concentration are also given. The elemental compositions of the prepared ceramics from EDS analysis, parameters of dielectric, ferroelectric and piezoelectric measurements, bond lengths and mid-bond electron density values from MEM analysis are also presented.

Keywords: X-Ray Diffraction, Scanning Electron Microscopy, EDS, Optical Studies, Dielectric Constant, Ferro Electricity, Piezo Electric Constant, D33, Charge Density

Contents

Materials Research Forum LLC
doi: http://dx.doi.org/10.21741/9781945291951

3.1 Introduction

In the work reported in this book, five different lead-free solid solutions have been synthesized by solid-state reaction method. The synthesized lead-fee solid solutions are as follows;

(i) $(1-x)(Na_{1-y}K_y)NbO_3-xBaTiO_3$ (x=0.1, 0.2; y=0.01, 0.05) (NKN-BT)

(ii) $(1-x)(Na_{1-y}K_y)(Nb_{1-z}Sb_z)O_3-xBaTiO_3$ (x=0.1, 0.2; y=0.03, 0.05; z=0.05, 0.1) (NKNS-BT)

(iii) $(1-x)(Na_{0.5}Bi_{0.5})TiO_3-xBaTiO_3$ (x=0.00, 0.04, 0.08, 0.12) (NBT-BT)

(iv) $(1-x)(K_{0.5}Bi_{0.5})TiO_3-xBaTiO_3$ (x=0.00, 0.08, 0.12) (KBT-BT)

(v) $(1-x)Ba(Zr_{0.2}Ti_{0.8})O_3-x(Ba_{0.7}Ca_{0.3})TiO_3$ (x=0.4, 0.5, 0.6) (BZT-BCT)

The results obtained from various characterization techniques and analytical methods have been presented in this chapter. The synthesized lead-free ceramics have been stucturally characterized by powder X-ray diffraction method (PXRD). The experimental X-ray diffraction data sets have been refined using Rietveld method (Rietveld, 1969) through JANA 2006 (Petříček *et al.*, 2006) software. The Rietveld refinement technique (Rietveld, 1969) refines various parameters such as lattice parameters, fractional coordinates, *etc.* The surface morphology of sintered samples has been analyzed through scanning electron microscopy (SEM). The elemental compositions of the solid solution samples have been analyzed by energy dispersive X-ray spectroscopy (EDS). The optical properties of the lead-free solid solutions have been investigated by UV-visible spectroscopy. Band gap values of prepared samples have been evaluated extrapolating the linear portion of the Tauc plot.

The dielectric constant (ε) of the ceramics as a function of temperature has been measured at different frequencies. The dielectric measurements revealed the structural phase transition of the solid solutions. The ferroelectric behaviour of the solid solutions has been analyzed from hysteresis loops measured at room temperature of polarization versus electric field. The piezoelectric constant (d_{33}) has been measured for the electrically poled samples at high electric field by a d_{33} meter. For the synthesized lead-free ceramics, the charge density distribution studies have been made using maximum entropy method (MEM) (Collins, 1982), employing PRIMA (Momma and Izumi, 2011) software.

3.2   Structural characterization - Powder X-ray diffraction

The phase purity and crystal structure of the prepared lead-free ceramics were confirmed by powder X-ray diffraction method (PXRD). The X-ray diffraction patterns of all lead-free ceramics were recorded at room temperature using CuKα radiation (λ=1.54056 Å) in the 2θ range of 10°-120° with a step size of 0.02°. The crystal structures of the samples were refined by the Rietveld refinement (Rietveld, 1969) technique using JANA 2006 (Petříček *et al.*, 2006) software. The Rietveld refinement (Rietveld, 1969) was used to minimize the difference between the calculated and observed XRD profiles. The experimental points are given as '××' and theoretical data are shown as solid lines in the figures of refined X-ray powder profiles. The difference between calculated and observed profile data is shown at the bottom along with Bragg's positions. The fitted XRD profiles for the prepared lead-free ceramics are presented in figures 3.1 to 3.9. The variations in the lattice parameters, cell volume, R_P (profile reliability index), R_{obs} (observed reliability index) and GOF (goodness of fit) are presented in Tables 3.1 to 3.5.

3.2.1   (1-x)(Na$_{1-y}$K$_y$)NbO$_3$-xBaTiO$_3$, (x=0.1, 0.2; y=0.01, 0.05)

The X-ray diffraction patterns of (1-x)(Na$_{1-y}$K$_y$)NbO$_3$-xBaTiO$_3$, (x=0.1, 0.2; y=0.01, 0.05) ceramics are shown in figures 3.1 (a)-(c). Figures 3.1 (b) and (c) show enlarged view of XRD peaks of (110) and (200) in the range of 2θ=31°-33° and 44°-47°. The profiles fitted for the prepared solid solutions using JANA 2006 (Petříček *et al.*, 2006) software are presented in figures 3.2 (a)-(d). The structural parameters, reliability indices and goodness of fit (GOF) are listed in Table 3.1.

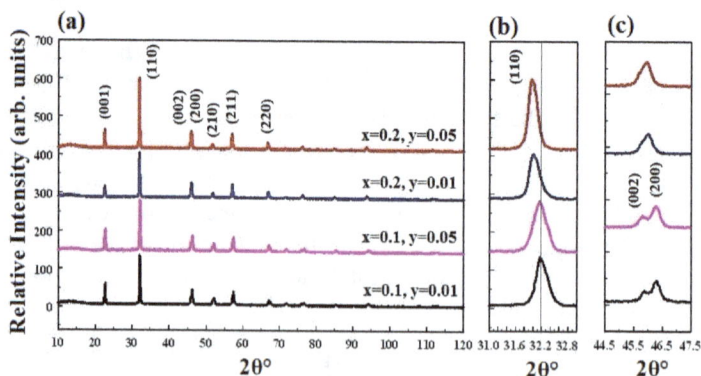

***Figure 3.1 (a)** X-ray diffraction patterns of (1-x)(Na$_{1-y}$K$_y$)NbO$_3$-xBaTiO$_3$, (x=0.1, 0.2; y=0.01, 0.05) ceramics **(b)** enlarged (110) peak & **(c)** enlarged (200) peak.*

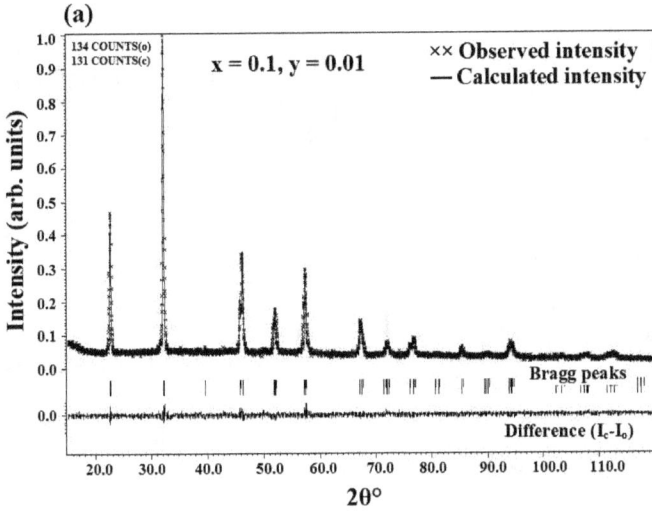

Figure 3.2 (a) *Fitted powder XRD profile for 0.9(Na$_{0.99}$K$_{0.01}$)NbO$_3$-0.1BaTiO$_3$.*

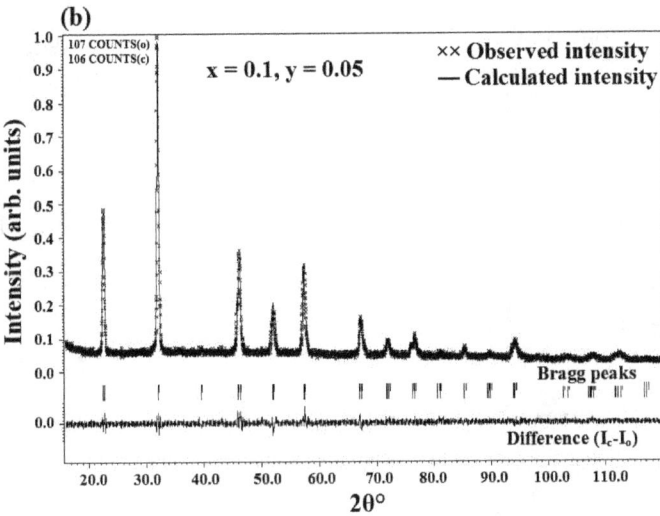

Figure 3.2 (b) *Fitted powder XRD profile for 0.9(Na$_{0.95}$K$_{0.05}$)NbO$_3$-0.1BaTiO$_3$.*

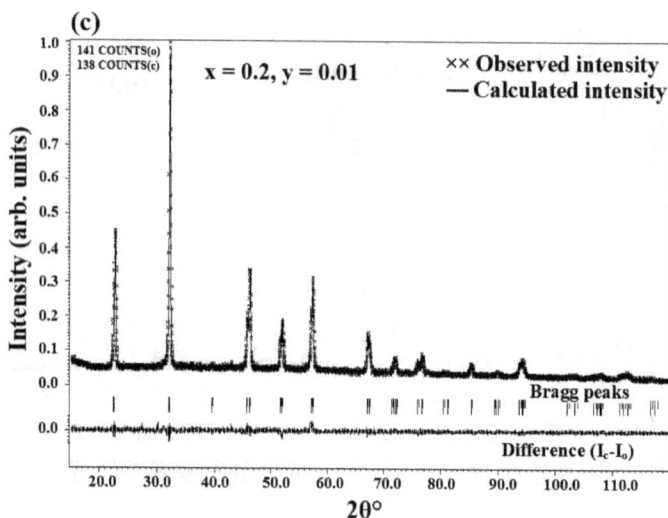

Figure 3.2 (c) Fitted powder XRD profile for $0.8(Na_{0.99}K_{0.01})NbO_3$-$0.2BaTiO_3$.

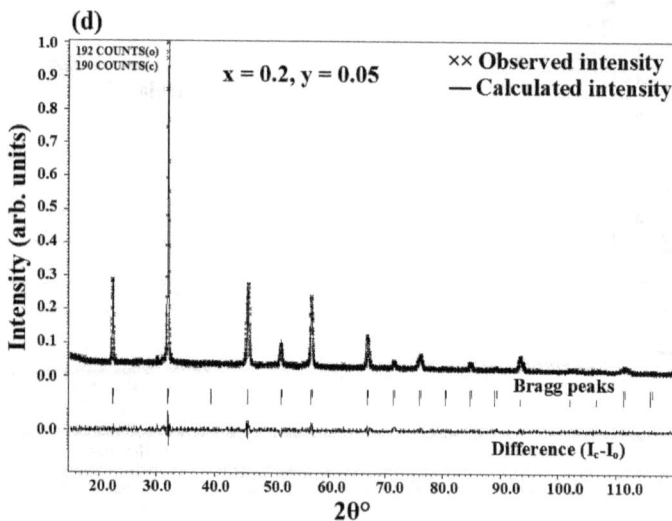

Figure 3.2 (d) Fitted powder XRD profile for $0.8(Na_{0.95}K_{0.05})NbO_3$-$0.2BaTiO_3$.

66

Materials Research Forum LLC
doi: http://dx.doi.org/10.21741/9781945291951

Table 3.1 *Refined structural parameters of (1-x)(Na$_{1-y}$K$_y$)NbO$_3$-xBaTiO$_3$ through refinement of powder XRD data.*

Parameter	x=0.1, y=0.01	x=0.1, y=0.05	x=0.2, y=0.01	x=0.2, y=0.05
a = b (Å)	3.9250 (4)	3.9271(5)	3.9562(6)	3.9613(11)
c (Å)	3.9587(13)	3.9642(4)	3.9672(6)	3.9763(11)
α=β=γ (°)	90	90	90	90
c/a	1.008	1.009	1.002	1.003
Unit cell Volume (Å^3)	60.99(4)	61.14(1)	61.43 (1)	62.16(2)
Density (gm/cc)	4.46(3)	4.47(1)	4.43 (1)	4.39(1)
R$_P$ (%)	5.97	5.74	7.06	6.34
R$_{obs}$ (%)	2.12	1.96	3.49	1.74
GOF	0.20	0.20	0.22	0.22
F$_{(000)}$	76	76	76	76

R$_P$ - Reliability index for profile

R$_{obs}$ - Reliability index for observed structure factors

GOF - Goodness of fit

F$_{(000)}$ - Number of electrons in the unit cell

3.2.2 (1-x)(Na$_{1-y}$K$_y$)(Nb$_{1-z}$Sb$_z$)O$_3$-xBaTiO$_3$, (x=0.1, 0.2; y=0.03, 0.05; z=0.05,0.1)

The X-ray diffraction patterns of (1-x)(Na$_{1-y}$K$_y$)(Nb$_{1-z}$Sb$_z$)O$_3$-xBaTiO$_3$, (x=0.1, 0.2; y=0.03, 0.05; z=0.05, 0.1) ceramics are given in figure 3.3 (a). Figure 3.3 (b) shows the enlarged XRD profiles of the (hkl) peak in 2θ range of 30°-34°. The crystal structure of the samples was refined using Rietveld method (Rietveld, 1969) and the fitted profiles are shown in figures 3.4 (a)-(d). The cell parameters and reliability factors obtained from Rietveld refinement (Rietveld, 1969) are listed in Table 3.2.

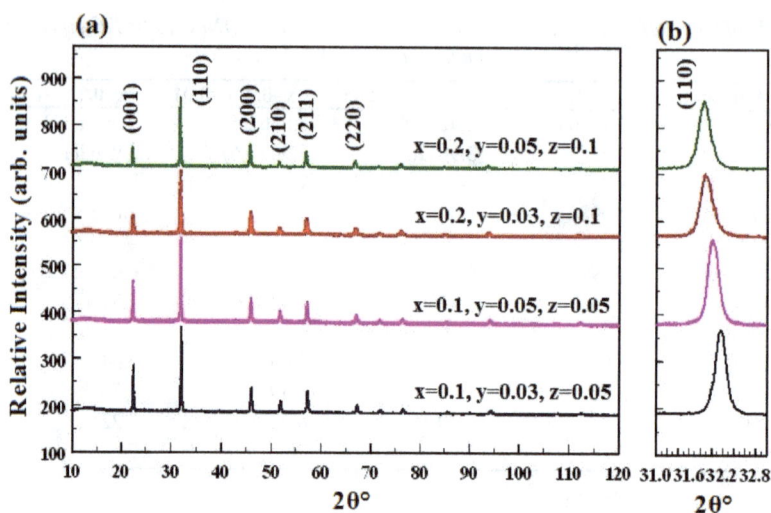

Figure 3.3 (a) *Powder XRD patterns of (1-x)(Na$_{1-y}$K$_y$)(Nb$_{1-z}$Sb$_z$)O$_3$-xBaTiO$_3$, (x=0.1, 0.2; y=0.03, 0.05; z=0.05, 0.1) ceramics & **(b)** Enlarged (110) peak.*

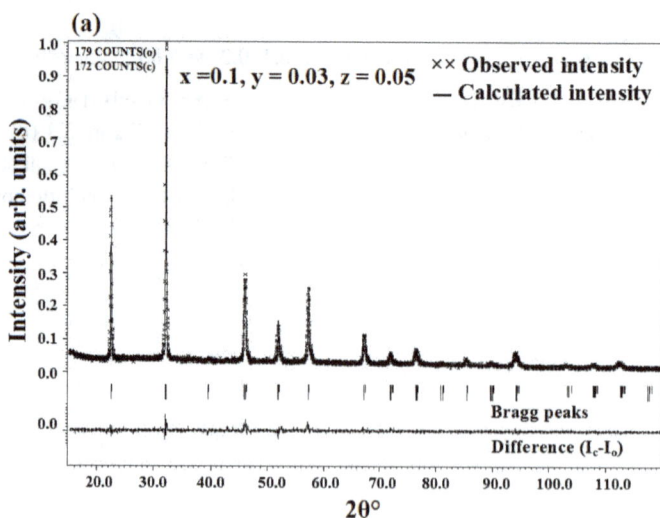

Figure 3.4 (a) *Fitted powder XRD profile for 0.9(Na$_{0.97}$K$_{0.03}$)(Nb$_{0.95}$Sb$_{0.05}$)O$_3$-0.1BaTiO$_3$.*

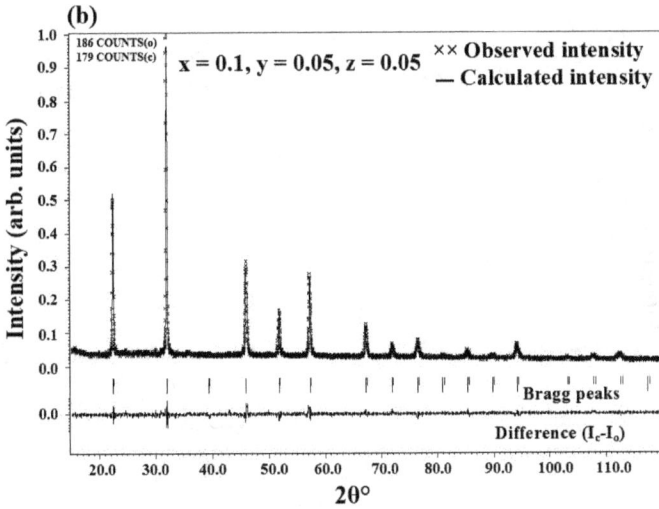

Figure 3.4 (b) Fitted powder XRD profile for 0.9(Na$_{0.95}$K$_{0.05}$)(Nb$_{0.95}$Sb$_{0.05}$)O$_3$-0.1BaTiO$_3$.

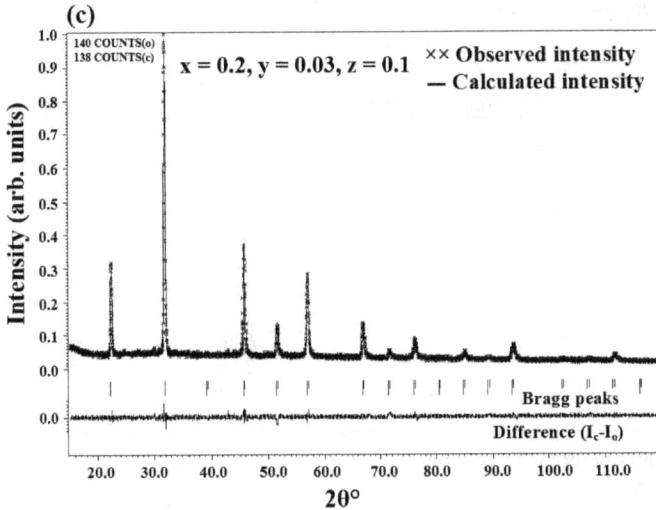

Figure 3.4 (c) Fitted powder XRD profile for 0.8(Na$_{0.97}$K$_{0.03}$)(Nb$_{0.9}$Sb$_{0.1}$)O$_3$-0.2BaTiO$_3$.

Materials Research Forum LLC
doi: http://dx.doi.org/10.21741/9781945291951

(d)

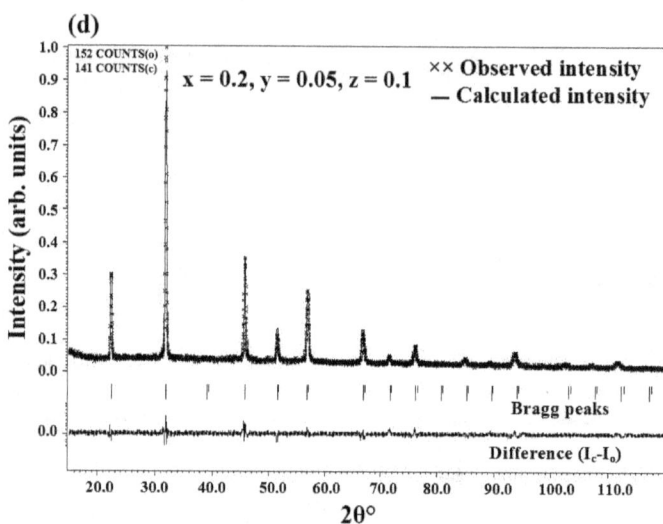

Figure 3.4 (d) *Fitted powder XRD profile for 0.8($Na_{0.95}K_{0.05}$)($Nb_{0.9}Sb_{0.1}$)O_3-0.2BaTiO$_3$.*

Table 3.2 *Refined structural parameters of (1-x)($Na_{1-y}K_y$)($Nb_{1-z}Sb_z$)O_3-xBaTiO$_3$ through refinement of powder XRD data.*

Parameter	x=0.1, y=0.03, z=0.05	x=0.1, y=0.05, z=0.05	x=0.2, y=0.03, z=0.1	x=0.2, y=0.05, z=0.1
a = b (Å)	3.9286(6)	3.9339(8)	3.9389(3)	3.9442(5)
c (Å)	3.9387(6)	3.9420(8)	3.9389(3)	3.9442(5)
α=β=γ (°)	90	90	90	90
c/a ratio	1.002	1.002	1	1
Unit cell Volume (Å^3)	60.79	60.99	61.11	61.35
Density (gm/cc)	4.48	4.52	4.45	4.44
R$_P$ (%)	6.32	6.74	6.56	7.67
R$_{obs}$ (%)	2.04	3.79	2.58	5.49
GOF	0.22	0.22	0.21	0.24
F$_{(000)}$	76	77	76	76

R$_P$ - Reliability index for profile

R$_{obs}$ - Reliability index for observed structure factors

GOF - Goodness of fit

F$_{(000)}$ - Number of electrons in the unit cell

3.2.3 (1-x)(Na$_{0.5}$Bi$_{0.5}$)TiO$_3$-xBaTiO$_3$, (x=0.00, 0.04, 0.08, 0.12)

The XRD patterns of (1-x)(Na$_{0.5}$Bi$_{0.5}$)TiO$_3$-xBaTiO$_3$, (x=0.00, 0.04, 0.08, 0.12) ceramics are presented in figures 3.5 (a) and (b). Figure 3.5 (b) shows the peak splitting of (002) and (200) peaks, which is attributed to phase transition to tetragonal system in (1-x)NBT-xBT ceramics from rhombohedral system. The fitted profiles of (1-x)NBT-xBT are shown in figures 3.6 (a)-(d). The reliability indices such as R$_{obs}$, R$_P$ and goodness of fit (GOF) and refined lattice parameters indicate good fit between the refined and observed profiles as given in Table 3.3.

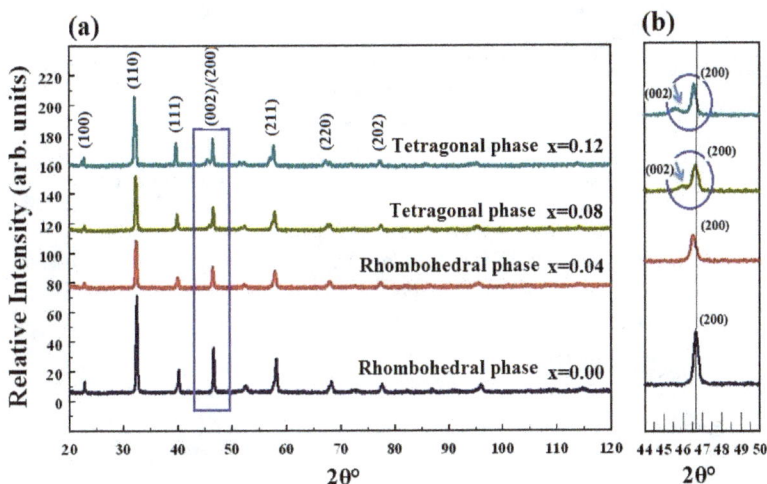

Figure 3.5 (a) *X-ray diffraction patterns of (1-x)(Na$_{0.5}$Bi$_{0.5}$)TiO$_3$-xBaTiO$_3$, (x=0.00, 0.04, 0.08, 0.12) ceramics & **(b)** Enlarged (200) peak (the circle represents peaks splitting of (002) and (200)).*

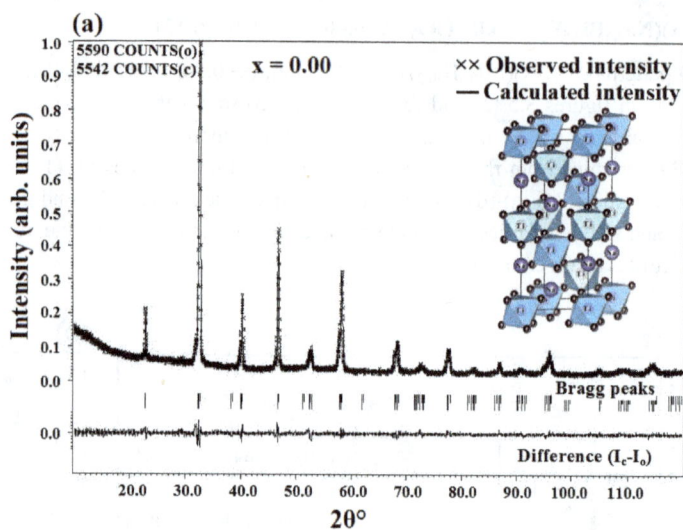

Figure 3.6 (a) Fitted powder XRD profile for $(Na_{0.5}Bi_{0.5})TiO_3$ (inset shows the structural model).

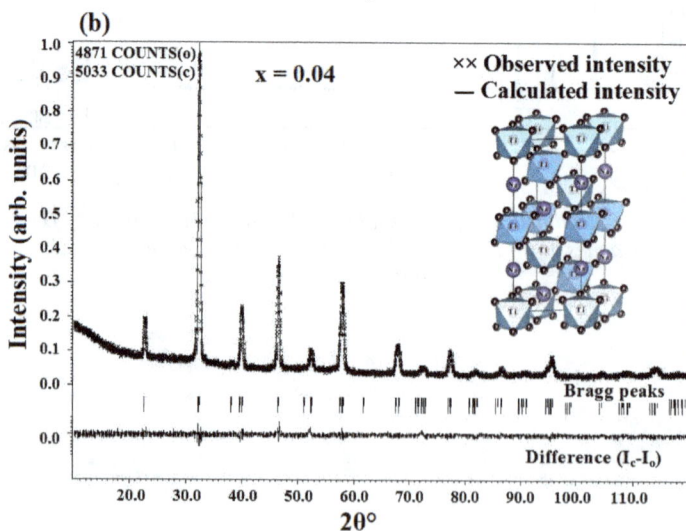

Figure 3.6 (b) Fitted powder XRD profile for $0.96(Na_{0.5}Bi_{0.5})TiO_3$-$0.04BaTiO_3$ (inset shows the structural model).

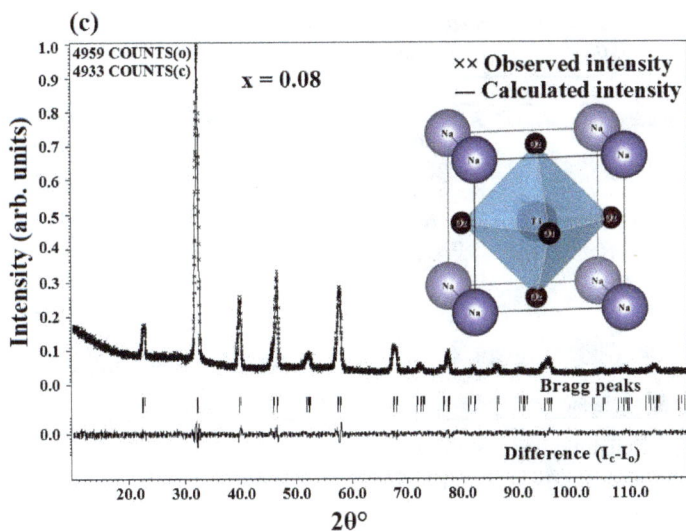

Figure 3.6 (c) *Fitted powder XRD profile for 0.92(Na$_{0.5}$Bi$_{0.5}$)TiO$_3$-0.08BaTiO$_3$ (inset shows the structural model).*

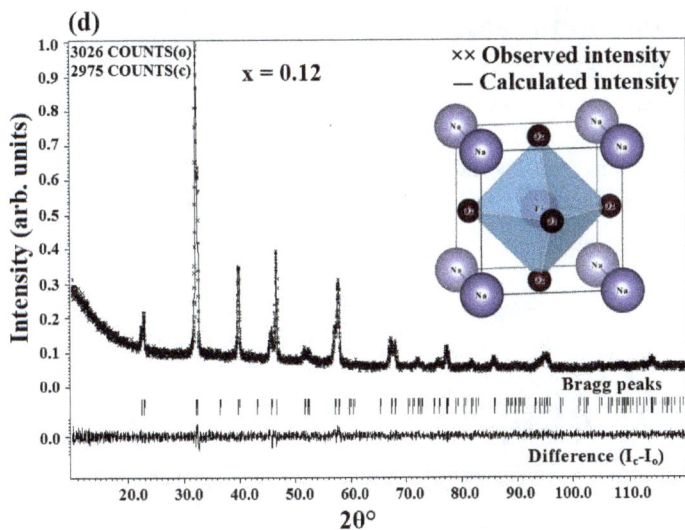

Figure 3.6 (d) *Fitted powder XRD profile for 0.88(Na$_{0.5}$Bi$_{0.5}$)TiO$_3$-0.12BaTiO$_3$ (inset shows the structural model).*

Table 3.3 *Refined structural parameters of (1-x)($Na_{0.5}Bi_{0.5}$)TiO_3-x$BaTiO_3$, (x=0.00, 0.04, 0.08, 0.12) phase through refinement of powder XRD data.*

Parameter	x=0.00	x=0.04	x=0.08	x=0.12
Space group	R3c	R3c	P4mm	P4mm
a = b (Å)	5.4746(9)	5.4809(16)	3.8746(12)	3.8879(2)
c (Å)	13.5207(8)	13.5322(14)	3.9226(9)	3.9582(2)
$\alpha = \beta \neq \gamma$ (°)	90°, 120°	90°, 120°	90°	90°
Unit cell Volume (Å³)	350.95(9)	352.06(2)	58.89(4)	59.83(6)
Density (gm/cc)	6.01(1)	5.99(3)	5.97(4)	5.87(6)
R_P (%)	4.83	5.40	4.98	4.89
R_{obs} (%)	1.12	0.70	2.27	3.18
GOF	0.16	0.15	0.15	0.16
$F_{(000)}$	558	558	93	93
Tolerance factor (t)	0.849	0.852	0.856	0.859

R_P - Reliability index for profile

R_{obs} - Reliability index for observed structure factors

GOF - Goodness of fit

$F_{(000)}$ - Number of electrons in the unit cell

3.2.4 (1-x)($K_{0.5}Bi_{0.5}$)TiO_3-x$BaTiO_3$, (x=0.00, 0.08, 0.12)

Figures 3.7 (a) and (b) show the powder XRD patterns of (1-x)($K_{0.5}Bi_{0.5}$)TiO_3-x$BaTiO_3$, (x=0.00, 0.08, 0.12) ceramics sintered at 1050°C for 3 h. The fitted profiles of XRD data sets for (1-x)($K_{0.5}Bi_{0.5}$)TiO_3-x$BaTiO_3$, (x=0.00, 0.08, 0.12) are shown in figures 3.8 (a)-(c). The reliability factors and goodness of fit obtained from Rietveld refinement (Rietveld, 1969) are given in Table 3.4.

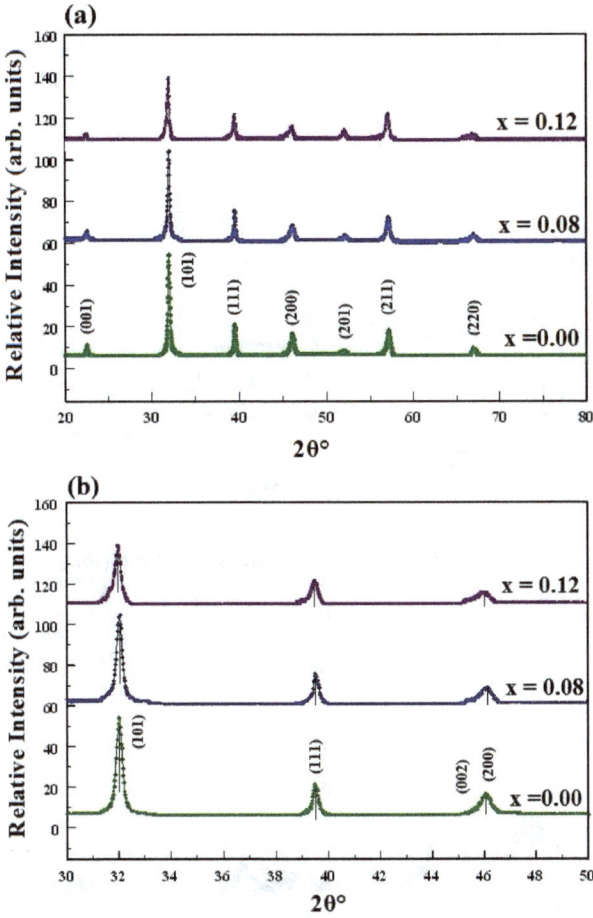

Figure 3.7 (a) *X-ray diffraction patterns of (1-x)(K$_{0.5}$Bi$_{0.5}$)TiO$_3$-xBaTiO$_3$, (x=0.00, 0.08, 0.12) ceramics at room temperature &* ***(b)*** *Enlarged XRD peaks of (101), (111) and (200).*

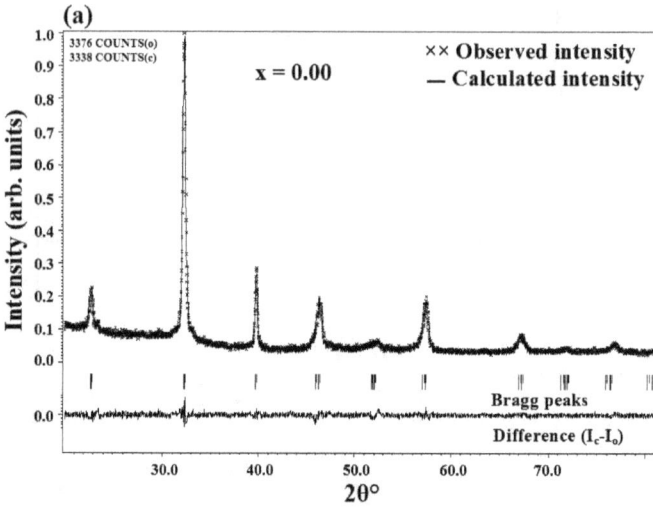

(a)

3376 COUNTS(o)
3338 COUNTS(c)

x = 0.00

×× **Observed intensity**
— **Calculated intensity**

Intensity (arb. units)

Bragg peaks

Difference (I_c-I_o)

$2\theta°$

Figure 3.8 (a) *Refined powder XRD profile for* $K_{0.5}Bi_{0.5}TiO_3$.

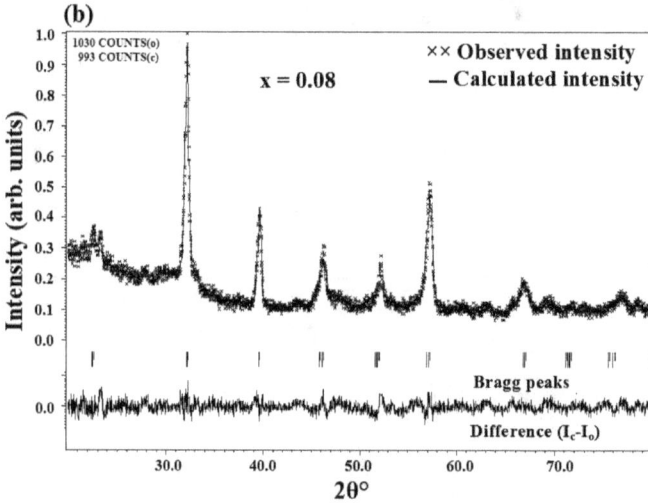

(b)

1030 COUNTS(o)
993 COUNTS(c)

x = 0.08

×× **Observed intensity**
— **Calculated intensity**

Intensity (arb. units)

Bragg peaks

Difference (I_c-I_o)

$2\theta°$

Figure 3.8 (b) *Refined powder XRD profile for* $0.92(K_{0.5}Bi_{0.5})TiO_3$-$0.08BaTiO_3$.

Materials Research Forum LLC
doi: http://dx.doi.org/10.21741/9781945291951

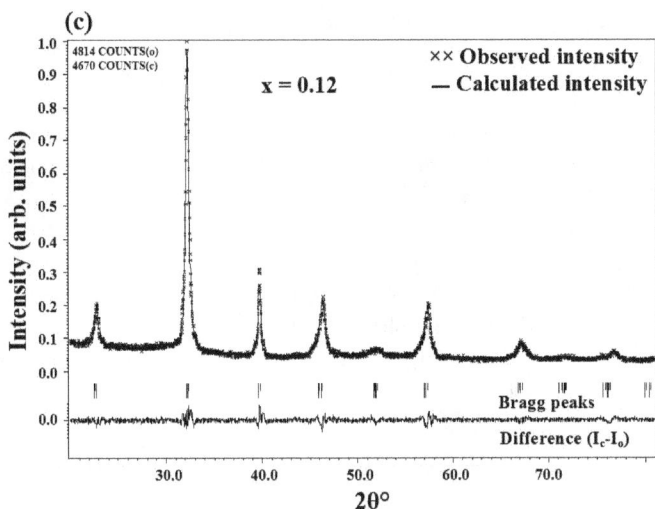

Figure 3.8 (c) Refined powder XRD profile for 0.88($K_{0.5}Bi_{0.5}$)TiO_3-0.12BaTiO_3.

Table 3.4 *Refined structural parameters of (1-x)($K_{0.5}Bi_{0.5}$)TiO_3-xBaTiO_3, (x=0.00, 0.08, 0.12) ceramics through refinement of powder XRD data*

Parameter	x=0.00	x=0.08	x=0.12
a = b (Å)	3.9760(6)	4.0139(3)	4.0287(2)
c (Å)	4.0018(6)	4.0201(3)	4.0541(2)
α=β=γ (°)	90	90	90
Unit cell Volume (Å³)	63.23(9)	64.79(6)	65.35(4)
Density (gm/cc)	5.77(1)	5.63(1)	5.58(3)
R_P (%)	5.90	5.88	6.06
R_{obs} (%)	2.83	5.66	4.80
GOF	1.14	1.68	1.44
$F_{(000)}$	97	97	97
c/a	1.006	1.001	1.006

R_P - Reliability index for profile
R_{obs} - Reliability index for observed structure factors
GOF - Goodness of fit
$F_{(000)}$ - Number of electrons in the unit cell

77

3.2.5 $(1-x)Ba(Zr_{0.2}Ti_{0.8})O_3-x(Ba_{0.7}Ca_{0.3})TiO_3$, (x=0.4, 0.5, 0.6)

Figures 3.9 (a) and (b) show the X-ray diffraction patterns of lead-free $(1-x)Ba(Zr_{0.2}Ti_{0.8})O_3-x(Ba_{0.7}Ca_{0.3})TiO_3$, (x=0.4, 0.5, 0.6) ceramics recorded at room temperature. Figure 3.9 (b) shows the enlarged (200) peak in the 2θ range $43°-47°$. Splitting of the (200) diffraction peak at $2\theta \sim 45°$ is observed in the sample of x=0.5 as shown in figure 3.9 (b). Figures 3.10 (a)-(c) show the fitted profiles of (1-x)BZT-xBCT, (x=0.4, 0.5, 0.6) ceramics with corresponding model of tetragonal structures respectively. Figure 3.10 (b) shows the mixed phase refinement profile for the morphotropic phase boundary (MPB) composition 0.5BZT-0.5BCT with a predominant tetragonal phase (P4mm) and a weak rhombohedral phase (R3m). Table 3.5 shows the values of lattice parameters, reliability indices and goodness of fit obtained from Rietveld refinement (Rietveld, 1969) technique.

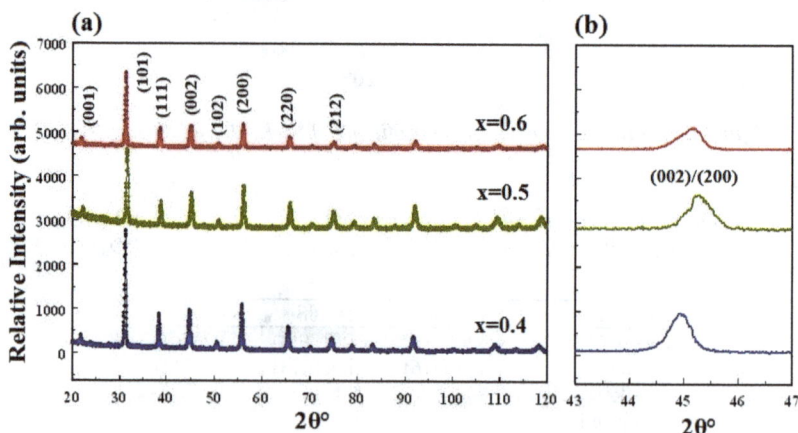

Figure 3.9 (a) *X-ray diffraction patterns of $(1-x)Ba(Zr_{0.2}Ti_{0.8})O_3-x(Ba_{0.7}Ca_{0.3})TiO_3$, (x=0.4, 0.5, 0.6) ceramics &* **(b)** *Expanded X-ray diffraction patterns for (002) peak*

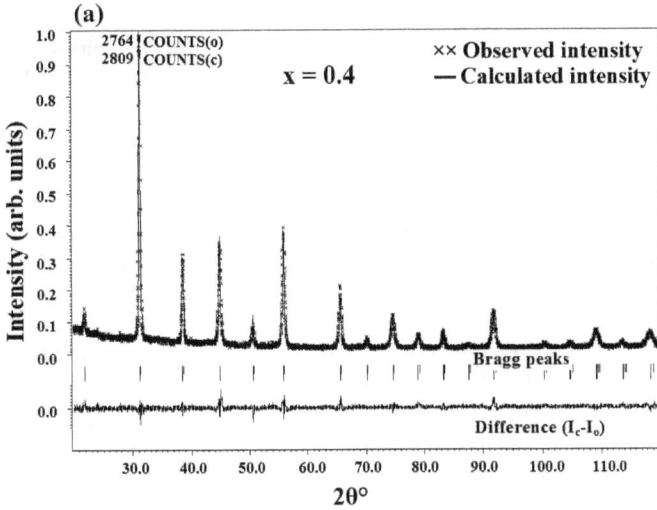

Figure 3.10 (a) *Fitted powder XRD profile for $0.6Ba(Zr_{0.2}Ti_{0.8})O_3$-$0.4(Ba_{0.7}Ca_{0.3})TiO_3$*

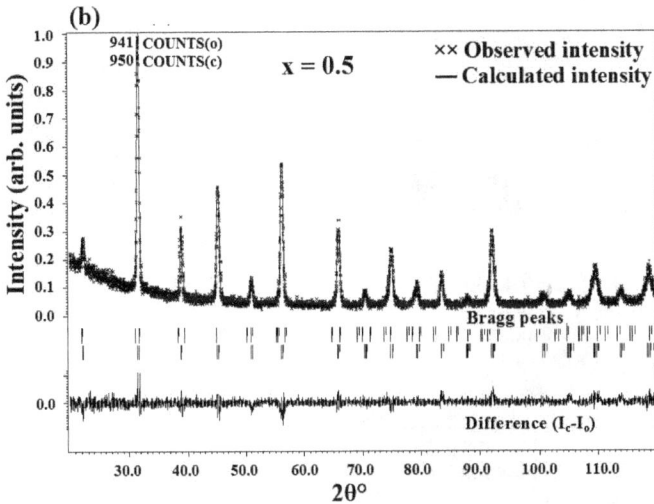

Figure 3.10 (b) *Fitted powder XRD profile for $0.5Ba(Zr_{0.2}Ti_{0.8})O_3$-$0.5(Ba_{0.7}Ca_{0.3})TiO_3$*

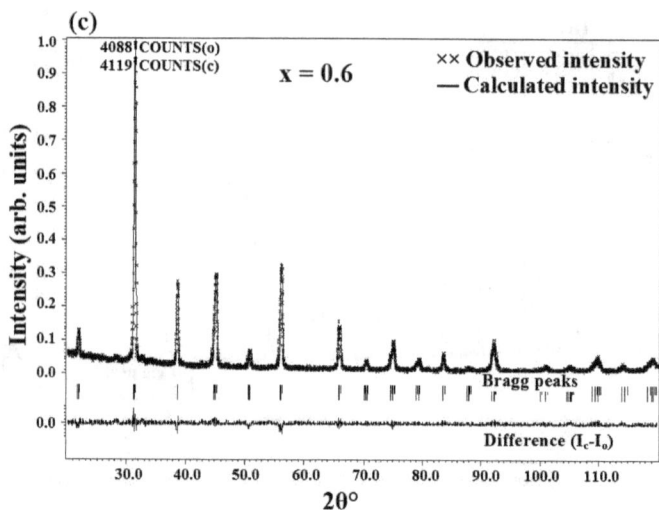

Figure 3.10 (c) *Fitted powder XRD profile for 0.4Ba(Zr$_{0.2}$Ti$_{0.8}$)O$_3$-0.6(Ba$_{0.7}$Ca$_{0.3}$)TiO$_3$*

Table 3.5 *Refined structural parameters for (1-x)Ba(Zr$_{0.2}$Ti$_{0.8}$)O$_3$-x(Ba$_{0.7}$Ca$_{0.3}$)TiO$_3$, (x=0.4, 0.5, 0.6) ceramics through refinement of powder XRD data*

Parameter	x=0.4	x=0.5		x=0.6
Phase	Tetragonal	Tetragonal	Rhombohedral	Tetragonal
Space group	P4mm	P4mm	R3m	P4mm
a = b (Å)	4.0040(1)	4.0141(1)	5.4455(2)	3.9843(5)
c (Å)	4.0065 (2)	3.9945(1)	6.6334(3)	4.0097(4)
Unit cell Volume (Å^3)	64.23(3)	64.36(3)	196.11(2)	63.65(1)
Density (gm/cc)	6.02	6.01	5.92	6.08
R$_P$ (%)	8.65	9.99	9.99	7.54
R$_{obs}$ (%)	4.31	5.02	9.61	3.49
GOF	1.24	1.13	1.13	1.17
F$_{(000)}$	102	102	306	102

R$_P$ - Reliability index for profile
R$_{obs}$ - Reliability index for observed structure factors
GOF - Goodness of fit
F$_{(000)}$ - Number of electrons in the unit cell

Materials Research Forum LLC
doi: http://dx.doi.org/10.21741/9781945291951

3.3 Microstructure and elemental characterization (SEM/EDS)

3.3.1 $(1-x)(Na_{1-y}K_y)NbO_3-xBaTiO_3$, (x=0.1, 0.2; y=0.01, 0.05)

Figures 3.11 (a)-(d) show the scanning electron micrographs of lead-free $(1-x)(Na_{1-y}K_y)NbO_3-xBaTiO_3$, (x=0.1, 0.2; y=0.01, 0.05) ceramics measured on the surface of the pellets. It can be seen that the $(1-x)(Na_{1-y}K_y)NbO_3-xBaTiO_3$, (x=0.1, 0.2; y=0.01, 0.05) particles are finely distributed without much agglomeration. The average particle size for the grown samples is given in Table 3.6.

Figures 3.12 (a)-(d) show the EDS spectra of $(1-x)(Na_{1-y}K_y)NbO_3-xBaTiO_3$, (x=0.1, 0.2; y=0.01, 0.05) ceramics. EDS measurement shows that all the elemental peaks corresponding to element of the prepared sample were present. The numerical values of atomic and weight percentages of the constituent elements are given in Table 3.7 (a) and (b).

(a)

(b)

(c)

(d)

Figure 3.11 *Scanning electron micrographs of $(1-x)(Na_{1-y}K_y)NbO_3-xBaTiO_3$,*
(a) x=0.1, y=0.01 (b) x=0.1, y=0.05 (c) x=0.2, y=0.01 & (d) x=0.2, y=0.05.

Table 3.6 *Particle size for $(1-x)(Na_{1-y}K_y)NbO_3-xBaTiO_3$, $(x=0.1, 0.2; y=0.01, 0.05)$ ceramics using SEM image.*

Sample concentration (x, y)	Average particle size (μm)
x=0.1, y=0.01	0.81
x=0.1, y=0.05	0.65
x=0.2, y=0.01	0.78
x=0.2, y=0.05	0.96

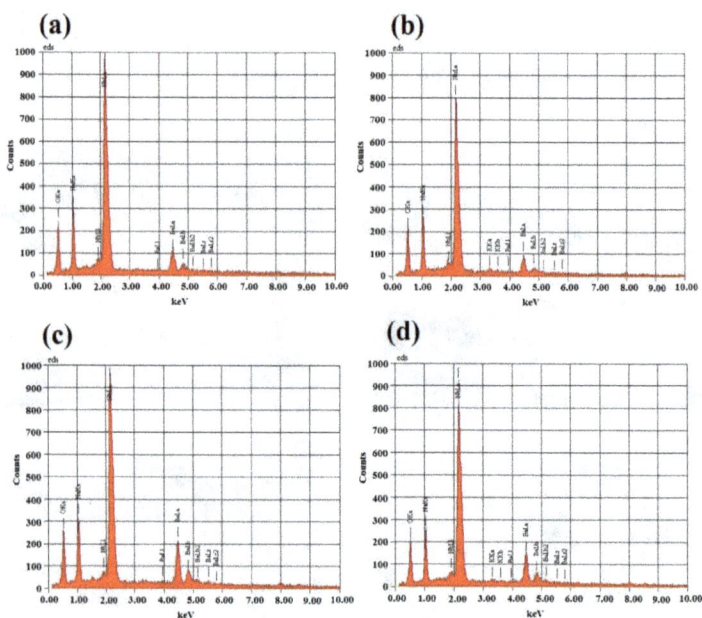

Figure 3.12 *EDS Spectra of $(1-x)(Na_{1-y}K_y)NbO_3-xBaTiO_3$, (a) x=0.1, y=0.01 (b) x=0.1, y=0.05 (c) x=0.2, y=0.01 & (d) x=0.2, y=0.05.*

Lead-free Piezo-Ceramic Solid Solutions, R. Saravanan   Materials Research Forum LLC
Materials Research Foundations **41** (2018)   doi: http://dx.doi.org/10.21741/9781945291951

Table 3.7 (a) *Elemental weight percentages of $(1-x)(Na_{1-y}K_y)NbO_3-xBaTiO_3$, (x=0.1, 0.2; y=0.01, 0.05) ceramics.*

Sample concentration	Weight (%)					
(x, y)	Na	K	Nb	Ba	Ti	O
x=0.1, y=0.01	13.13	0.52	46.03	11.13	1.28	27.91
x=0.1, y=0.05	13.06	0.77	47.79	9.37	1.51	28.51
x=0.2, y=0.01	12.99	0.41	39.63	20.08	3.11	23.78
x=0.2, y=0.05	13.63	0.65	47.97	23.60	3.67	11.10

Table 3.7 (b) *Elemental atomic percentages of $(1-x)(Na_{1-y}K_y)NbO_3-xBaTiO_3$, (x=0.1, 0.2; y=0.01, 0.05) ceramics.*

Sample concentration	Atomic (%)					
(x, y)	Na	K	Nb	Ba	Ti	O
x=0.1, y=0.01	19.58	0.66	16.84	2.75	0.91	59.30
x=0.1, y=0.05	18.87	1.36	17.41	2.31	1.07	60.33
x=0.2, y=0.01	21.10	0.75	13.63	5.60	2.43	56.51
x=0.2, y=0.05	29.81	2.45	25.96	6.66	3.86	31.26

3.3.2 $(1-x)(Na_{1-y}K_y)(Nb_{1-z}Sb_z)O_3-xBaTiO_3$, (x=0.1,0.2; y=0.03,0.05; z=0.05,0.1)

Figures 3.13 (a)-(d) display the SEM micrographs of the sintered samples of lead-free $(1-x)(Na_{1-y}K_y)(Nb_{1-z}Sb_z)O_3-xBaTiO_3$, (x=0.1, 0.2; y=0.03, 0.05; z=0.05, 0.1) ceramics. The SEM images of prepared pellet samples show rectangular-like grains. The average particle size for the grown samples is given in Table 3.8.

The compositional analysis was carried out through energy dispersive by X-ray spectra and is shown in figures 3.14 (a)-(d). The quantitative analysis of $(1-x)(Na_{1-y}K_y)(Nb_{1-z}Sb_z)O_3-xBaTiO_3$ obtained from the atomic and weight percentage values and the EDS results are listed in Tables 3.9 (a) and (b). The observed change in the weight percentage of Ba ions indicate that the Ba^{2+} incorporate into $Na_{1-y}K_yNb_{1-z}Sb_zO_3$.

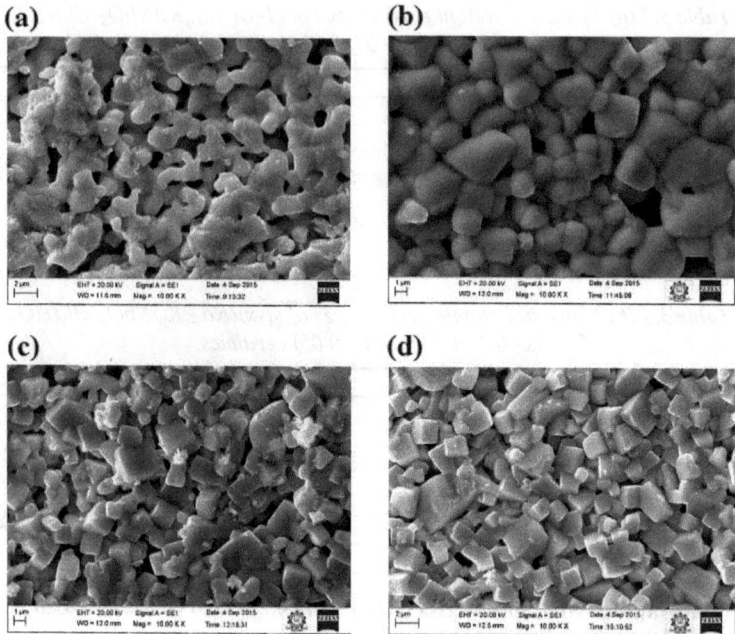

Figure 3.13 *Scanning electron microscopy images of $(1-x)(Na_{1-y}K_y)(Nb_{1-z}Sb_z)O_3$-$xBaTiO_3$, (a) x=0.1, y=0.03, z=0.05 (b) x=0.1, y=0.05, z=0.05 (c) x=0.2, y=0.03, z=0.1 & (d) x=0.2, y=0.05, z=0.1.*

Table 3.8 *Particle size for $(1-x)(Na_{1-y}K_y)(Nb_{1-z}Sb_z)O_3$-$xBaTiO_3$, (x=0.1, 0.2; y=0.03, 0.05; z=0.05, 0.1) ceramics using SEM image.*

Sample concentration (x, y, z)	Average particle size (µm)
x=0.1, y=0.03, z=0.05	1.5
x=0.1, y=0.05, z=0.05	0.53
x=0.2, y=0.03, z=0.1	0.46
x=0.2, y=0.05, z=0.1	0.96

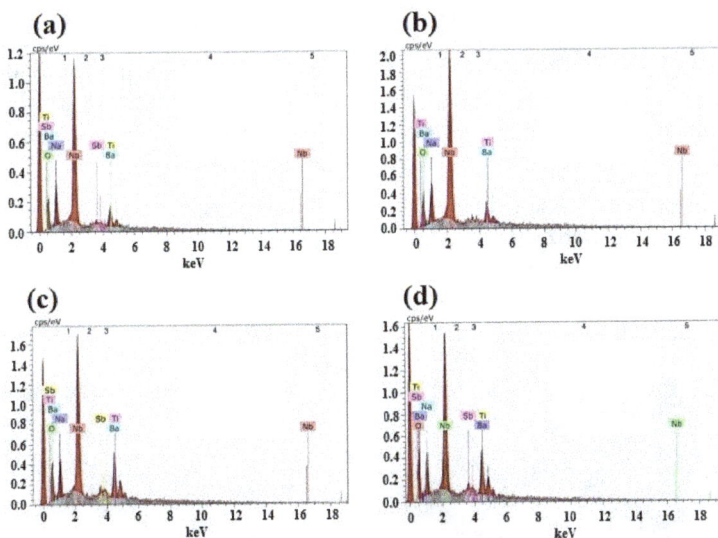

Figure 3.14 *EDS spectra of $(1-x)(Na_{1-y}K_y)(Nb_{1-z}Sb_z)O_3-xBaTiO_3$, (a) x=0.1, y=0.03, z=0.05 (b) x=0.1, y=0.05, z=0.05 (c) x=0.2, y=0.03, z=0.1 (d) x=0.2, y=0.05, z=0.1.*

Table 3.9 (a) *Elemental weight percentages of $(1-x)(Na_{1-y}K_y)(Nb_{1-z}Sb_z)O_3-xBaTiO_3$, (x=0.1, 0.2; y=0.03, 0.05; z=0.05, 0.1) ceramics.*

Sample concentration (x, y, z)	Weight (%)						
	Na	K	Nb	Sb	Ba	Ti	O
x=0.1, y=0.03, z=0.05	14.18	1.25	43.3	1.23	10.32	1.81	27.85
x=0.1, y=0.05, z=0.05	13.47	1.5	44.52	1.73	9.07	1.66	27.90
x=0.2, y=0.03, z=0.1	12.37	1.20	39.29	1.51	17.17	2.1	25.58
x=0.2, y=0.05, z=0.1	11.53	1.56	40.09	1.95	13.93	2.4	28.48

Table 3.9 (b) *Elemental atomic percentages of $(1-x)(Na_{1-y}K_y)(Nb_{1-z}Sb_z)O_3-xBaTiO_3$, (x=0.1, 0.2; y=0.03, 0.05; z=0.05, 0.1) ceramics*

Sample concentration (x, y, z)	Atomic (%)						
	Na	K	Nb	Sb	Ba	Ti	O
x=0.1, y=0.03, z=0.05	20.84	0.31	16.21	0.34	2.54	1.28	58.79
x=0.1, y=0.05, z=0.05	19.89	0.46	17.53	0.45	2.24	1.13	59.21
x=0.2, y=0.03, z=0.1	18.68	0.35	15.43	0.45	5.11	1.99	58.33
x=0.2, y=0.05, z=0.1	17.31	0.47	15.4	0.55	3.50	1.78	61.40

3.3.3 (1-x)(Na$_{0.5}$Bi$_{0.5}$)TiO$_3$-xBaTiO$_3$, (x=0.00, 0.04, 0.08, 0.12)

Figures 3.15 (a)-(c) show the SEM micrographs of (1-x)(Na$_{0.5}$Bi$_{0.5}$)TiO$_3$-xBaTiO$_3$, (x=0.00, 0.04, 0.08, 0.12) ceramics. The average particle size of the prepared samples was found to increase with the addition of Ba^{2+} concentration in the host lattice. The average particle size for the grown samples is listed in Table 3.10.

The distribution of each chemical element for the composition of (1-x)NBT-xBT ceramics was characterized by the element mapping of the selected cross section of the sample using energy dispersive X-ray spectroscopy (EDS) analysis. The EDS results are presented in figures 3.16 (a)-(d). The percentages of various elements present in the prepared solid solution samples are listed in Table 3.11.

(a) **(b)**

(c) **(d)**

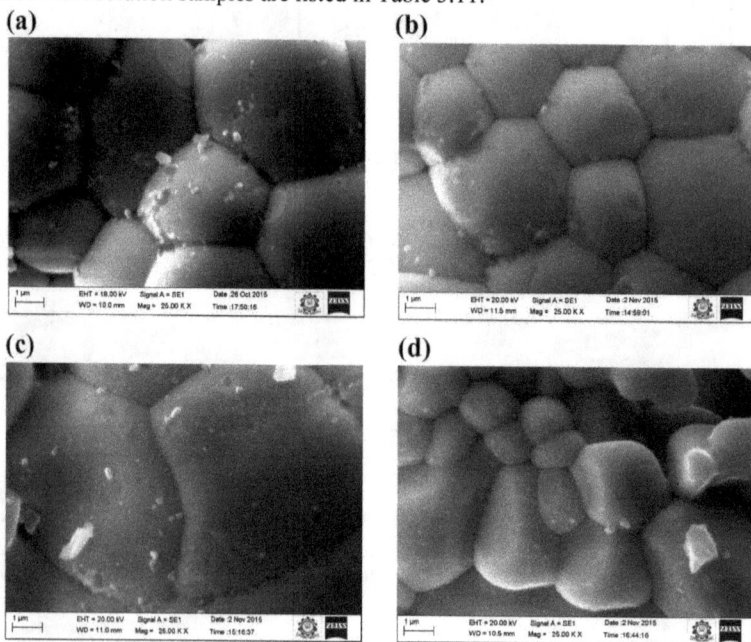

Figure 3.15 SEM micrographs of (1-x)(Na$_{0.5}$Bi$_{0.5}$)TiO$_3$-xBaTiO$_3$ ceramics,
(a) x=0.00 (b) x=0.04 (c) x=0.08 & (d) x=0.12.

Materials Research Forum LLC
doi: http://dx.doi.org/10.21741/9781945291951

Table 3.10 *Particle size for (1-x)(Na$_{0.5}$Bi$_{0.5}$)TiO$_3$-xBaTiO$_3$, (x=0.00, 0.04, 0.08, 0.12) ceramics using SEM image.*

Sample concentration (x)	Average particle size (µm)
x=0.00	1.53
x=0.04	1.85
x=0.08	3.20
x=0.12	1.30

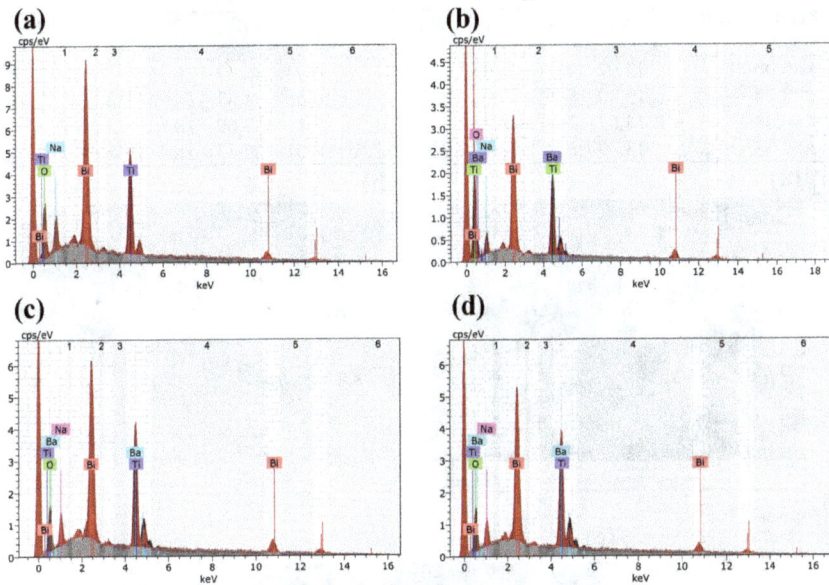

Figure 3.16 *EDS spectra (1-x)(Na$_{0.5}$Bi$_{0.5}$)TiO$_3$-xBaTiO$_3$ ceramics, **(a)** x=0.00 **(b)** x=0.04 **(c)** x=0.08 & **(d)** x=0.12*

3.3.4   (1-x)(K$_{0.5}$Bi$_{0.5}$)TiO$_3$-xBaTiO$_3$, (x=0.00, 0.08, 0.12)

The surface morphologies of (1-x)(K$_{0.5}$Bi$_{0.5}$)TiO$_3$-xBaTiO$_3$, (x=0.00, 0.08, 0.12) ceramics were analyzed by scanning electron microscopy (SEM). The SEM images were recorded with different magnifications (×1500, ×5000, ×10000). Figures 3.17 (a)-(c) show the SEM images of (1-x)(K$_{0.5}$Bi$_{0.5}$)TiO$_3$-xBaTiO$_3$ ceramics for the magnification ×5000. The average particle size for the prepared samples is given in Table 3.12.

The energy dispersive X-ray spectra indicate the presence of constituent element in the $(K_{0.5}Bi_{0.5})TiO_3$ lattice and the recorded EDS spectra are presented in figures 3.18 (a)-(c). The quantitative analysis revealed that the formula of the solid solution was $(1-x)(K_{0.5}Bi_{0.5})TiO_3$-$xBaTiO_3$. The observed atomic and weight percentages are given in Table 3.13.

Table 3.11 Elemental compositions of $(1-x)(Na_{0.5}Bi_{0.5})TiO_3$-$xBaTiO_3$, $(x=0.00, 0.04, 0.08, 0.12)$ ceramics determined by EDS analysis

Sample concentration (x)	Atomic percent (%)					Weight percent (%)				
	Na	Bi	Ti	Ba	O	Na	Bi	Ti	Ba	O
x=0.00	12.97	9.60	23.65	-	53.78	6.94	46.69	26.35	-	20.02
x=0.04	13.80	8.98	19.06	6.57	51.59	6.56	38.82	18.88	18.66	17.08
x=0.08	14.62	7.54	18.38	7.70	51.77	7.19	33.69	18.82	22.60	17.71
x=0.12	13.39	6.89	18.17	9.25	52.29	6.51	30.48	18.41	26.89	17.70

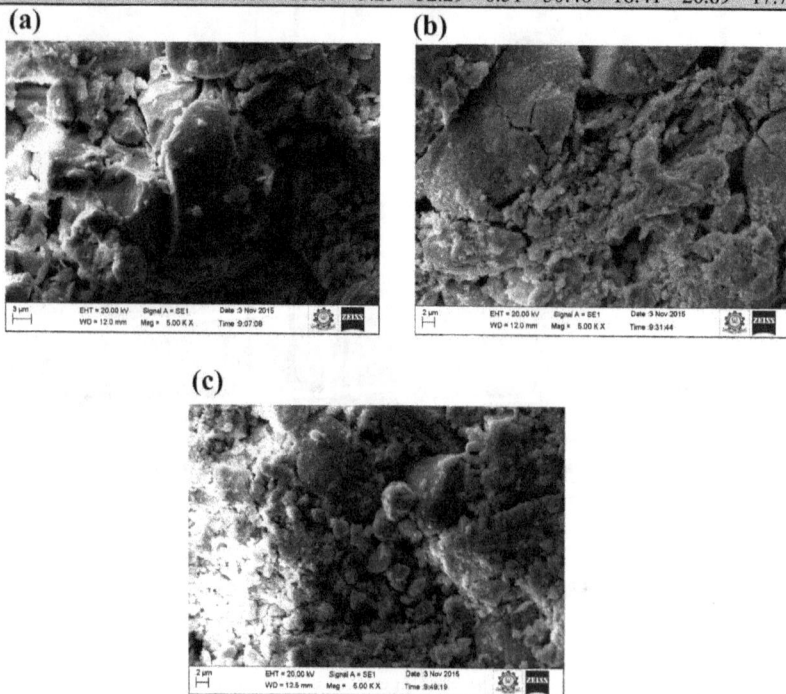

(a)

(b)

(c)

Figure 3.17 SEM images of $(1-x)(K_{0.5}Bi_{0.5})TiO_3$-$xBaTiO_3$ ceramics, **(a)** x=0.00 **(b)** x=0.08 & **(c)** x=0.12.

Lead-free Piezo-Ceramic Solid Solutions, R. Saravanan
Materials Research Foundations **41** (2018)

Materials Research Forum LLC
doi: http://dx.doi.org/10.21741/9781945291951

Table 3.12 *Particle size for $(1-x)(K_{0.5}Bi_{0.5})TiO_3$-$xBaTiO_3$, ($x=0.00$, 0.08, 0.12) ceramics using SEM image.*

Sample concentration (x)	Average particle size (µm)
x=0.00	0.81
x=0.08	0.61
x=0.12	0.43

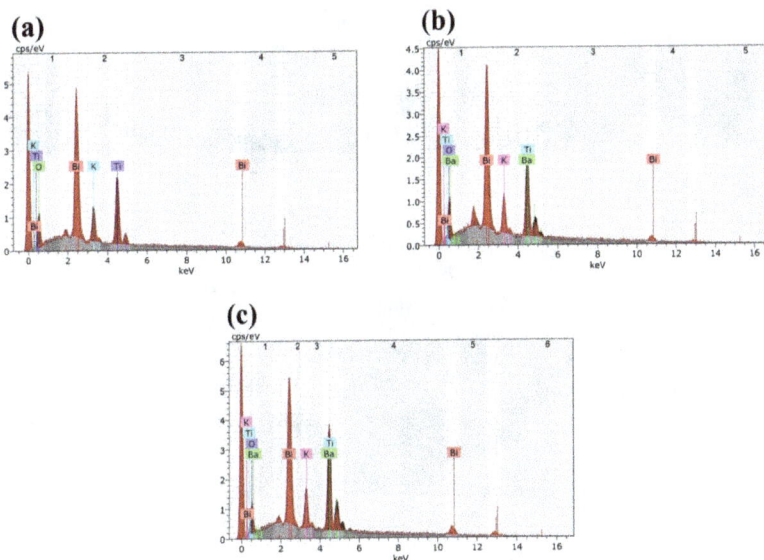

(a)

(b)

(c)

Figure 3.18 *EDS spectra $(1-x)(K_{0.5}Bi_{0.5})TiO_3$-$xBaTiO_3$ ceramics **(a)** $x=0.00$ **(b)** $x=0.08$ & **(c)** $x=0.12$.*

Table 3.13 *Elemental compositions of $(1-x)(K_{0.5}Bi_{0.5})TiO_3$-$xBaTiO_3$, ($x=0.00$, 0.08, 0.12) ceramics determined by EDS analysis.*

Sample concentration (x)	Atomic percent (%)					Weight percent (%)				
	K	Bi	Ti	Ba	O	K	Bi	Ti	Ba	O
x=0.00	9.12	6.32	26.67	-	5.89	9.20	34.02	32.91	-	23.87
x=0.08	8.76	5.06	21.67	8.39	56.12	7.63	23.57	23.12	25.68	20.00
x=0.12	9.06	7.43	17.80	12.78	52.92	6.61	28.97	15.89	32.74	15.79

3.3.5 $(1-x)Ba(Zr_{0.2}Ti_{0.8})O_3$-$x(Ba_{0.7}Ca_{0.3})TiO_3$, $(x=0.4, 0.5, 0.6)$

The SEM micrographs of $(1-x)Ba(Zr_{0.2}Ti_{0.8})O_3$-$x(Ba_{0.7}Ca_{0.3})TiO_3$, $(x=0.4, 0.5, 0.6)$ ceramics are shown in figures 3.19 (a)-(c). The particle size of $(1-x)Ba(Zr_{0.2}Ti_{0.8})O_3$-$x(Ba_{0.7}Ca_{0.3})TiO_3$, $(x=0.4, 0.5, 0.6)$ ceramics is listed in Table 3.14.

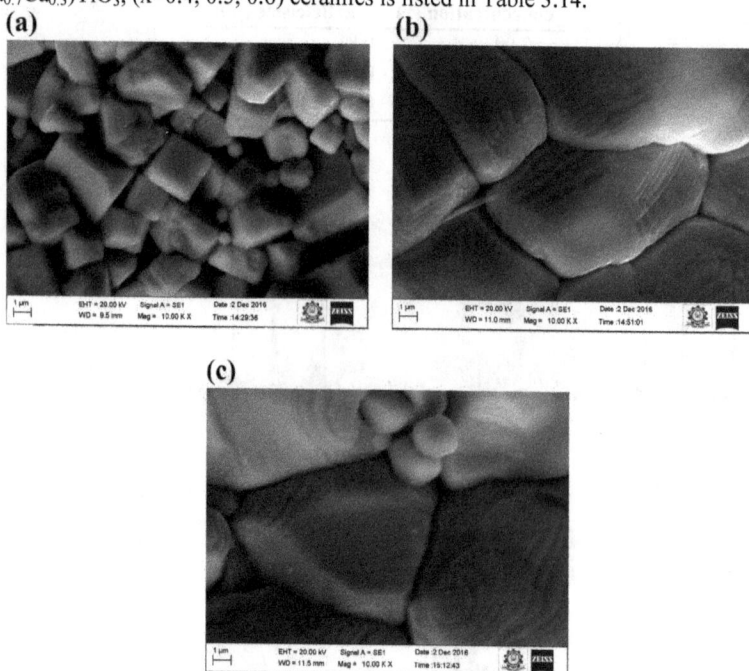

(a)

(b)

(c)

Figure 3.19 *SEM images of $(1-x)Ba(Zr_{0.2}Ti_{0.8})O_3$-$x(Ba_{0.7}Ca_{0.3})TiO_3$ ceramics, (a) $x=0.4$ (b) $x=0.5$ & (c) $x=0.6$.*

Table 3.14 *Particle size for $(1-x)Ba(Zr_{0.2}Ti_{0.8})O_3$-$x(Ba_{0.7}Ca_{0.3})TiO_3$, $(x=0.4, 0.5, 0.6)$ ceramics using SEM image.*

Sample concentration (x)	Average particle size (µm)
x=0.4	1.14
x=0.5	2.05
x=0.6	1.57

The energy dispersive X-ray spectra of $(1-x)Ba(Zr_{0.2}Ti_{0.8})O_3-x(Ba_{0.7}Ca_{0.3})TiO_3$, $(x=0.4$, 0.5, 0.6) ceramics were recorded to determine the chemical composition of the solid solution and are presented in figures 3.20 (a)-(c). The presence of all elements in these ceramics detected by EDS spectra are listed in Table 3.15.

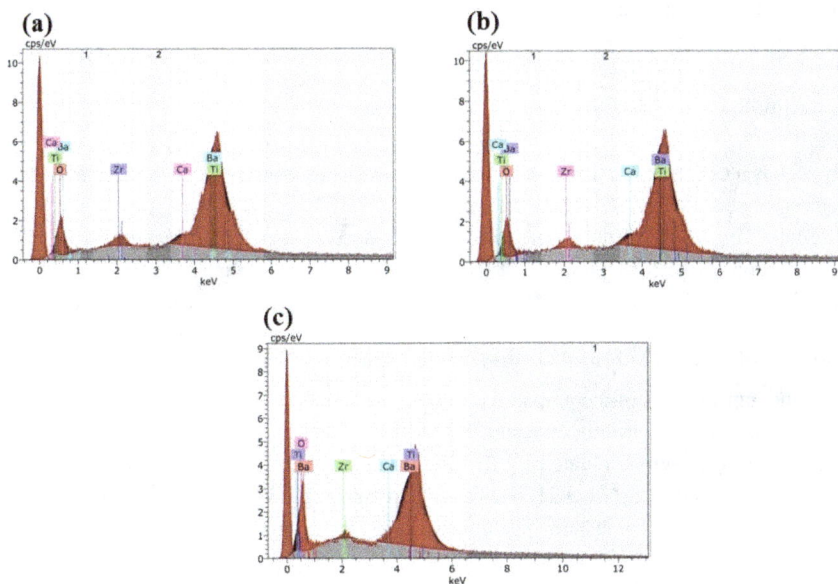

Figure 3.20 *EDS spectra of $(1-x)Ba(Zr_{0.2}Ti_{0.8})O_3-x(Ba_{0.7}Ca_{0.3})TiO_3$ ceramics, (a) $x=0.4$ (b) $x=0.5$ & (c) $x=0.6$.*

Table 3.15 *EDS elemental compositions of $(1-x)Ba(Zr_{0.2}Ti_{0.8})O_3-x(Ba_{0.7}Ca_{0.3})TiO_3$, $(x=0.4, 0.5, 0.6)$ ceramics determined by EDS analysis.*

Sample concentration (x)	Atomic (%)					Weight (%)				
	Ba	Zr	Ca	Ti	O	Ba	Zr	Ca	Ti	O
x=0.4	4.67	1.59	3.15	22.55	68.04	20.82	4.72	4.10	35.04	35.33
x=0.5	7.45	1.69	3.32	22.86	65.16	29.83	4.50	3.32	31.94	30.41
x=0.6	4.63	1.14	3.05	20.27	70.91	21.42	3.52	4.11	32.71	38.24

3.4   Optical characterization

The optical properties of the lead-free ceramics were analyzed by UV-visible spectroscopy. The optical band gap energy was estimated by the method proposed by Wood and Tauc (1972). According to Wood and Tauc (Wood and Tauc, 1972) equation, the optical band gap is associated with the absorbance and photon energy by the following equation.

$$\alpha h\nu = A(h\nu - E_g)^n \tag{3.1}$$

where A is a constant, α is the absorbance, h is the Planck constant, ν is the frequency, E_g is energy band gap, n=1/2 for direct band gap materials and n=2 for indirect band gap materials. The graph is drawn with energy (hν) value in x-axis and $(\alpha h\nu)^2$ in y-axis. The extrapolation of the linear portion of the curve is presented in figures 3.21 to 3.25. The optical band gap energy values of the lead-free ceramics are given in Tables 3.16 to 3.20.

3.4.1   (1-x)(Na$_{1-y}$K$_y$)NbO$_3$-xBaTiO$_3$, (x=0.1, 0.2; y=0.01, 0.05)

The band gap energies of the prepared (1-x)(Na$_{1-y}$K$_y$)NbO$_3$-xBaTiO$_3$, (x=0.1, 0.01, 0.05) ceramics were evaluated using UV-visible spectra. The optical band gaps of ceramic samples were evaluated by the Tauc relation (Wood and Tauc, 1972) and the plots are presented in figure 3.21. The band gap values of (1-x)(Na$_{1-y}$K$_y$)NbO$_3$-xBaTiO$_3$ ceramics are listed in Table 3.16. The optical band gap energy decreases with the increase of BaTiO$_3$ content.

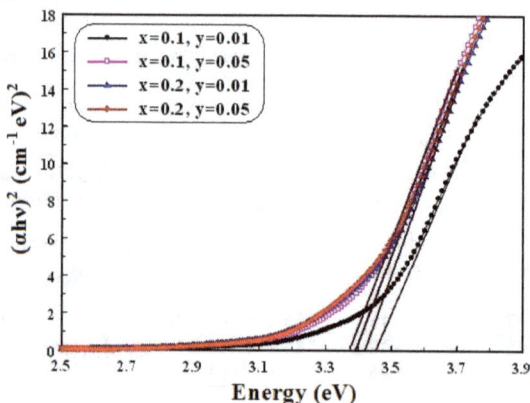

Figure 3.21 Tauc plot for (1-x)(Na$_{1-y}$K$_y$)NbO$_3$-xBaTiO$_3$, (x=0.1, 0.2; y=0.01, 0.05) ceramics.

Table 3.16 *Optical band gap values for (1-x)(Na$_{1-y}$K$_y$)NbO$_3$-xBaTiO$_3$, (x=0.1, 0.2; y=0.01, 0.05) ceramics from UV-visible analysis.*

Sample concentration (x)	Band gap (eV)
x=0.1, y=0.01	3.447
x=0.1, y=0.05	3.373
x=0.2, y=0.01	3.421
x=0.2, y=0.05	3.392

3.4.2 (1-x)(Na$_{1-y}$K$_y$)(Nb$_{1-z}$Sb$_z$)O$_3$-xBaTiO$_3$, (x=0.1,0.2; y=0.03,0.05; z=0.05,0.1)

The optical band gap values of (1-x)(Na$_{1-y}$K$_y$)(Nb$_{1-z}$Sb$_z$)O$_3$-xBaTiO$_3$, (x=0.1, 0.2; y=0.03, 0.05; z=0.05, 0.1) ceramics were determined using UV-visible spectroscopy. The optical band gaps of the prepared samples are associated with optical absorbance and photon energy, which were determined by Tauc equation (Wood and Tauc, 1972). The optical band gap E$_g$ was obtained from the plots of (hv) versus (αhv)^2 as shown in figure 3.22. The band gap values of (1-x)(Na$_{1-y}$K$_y$)(Nb$_{1-z}$Sb$_z$)O$_3$-xBaTiO$_3$ ceramics are given in Table 3.17.

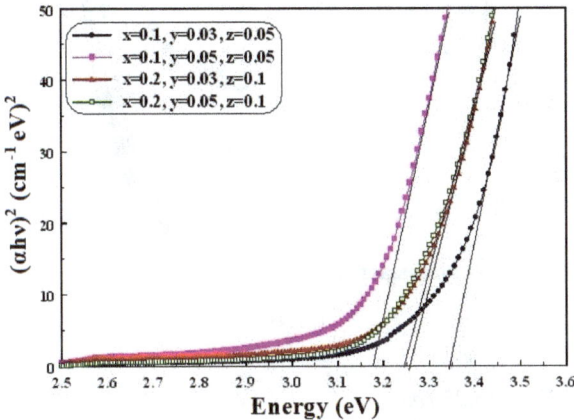

Figure 3.22 *Tauc plot of (1-x)(Na$_{1-y}$K$_y$)(Nb$_{1-z}$Sb$_z$)O$_3$-xBaTiO$_3$, (x=0.1, 0.2; y=0.03, 0.05; z=0.05, 0.1) ceramics.*

Table 3.17 *Optical band gap values for* $(1-x)(Na_{1-y}K_y)(Nb_{1-z}Sb_z)O_3-xBaTiO_3$, *(x=0.1, 0.2; y=0.03, 0.05; z=0.05, 0.1) ceramics from UV-visible analysis.*

Sample concentration (x)	Band gap (eV)
x=0.1, y=0.03, z=0.05	3.343
x=0.1, y=0.05, z=0.05	3.177
x=0.2, y=0.03, z=0.1	3.255
x=0.2, y=0.05, z=0.1	3.243

3.4.3 $(1-x)(Na_{0.5}Bi_{0.5})TiO_3-xBaTiO_3$, (x=0.00, 0.04, 0.08, 0.12)

The optical properties of $(1-x)(Na_{0.5}Bi_{0.5})TiO_3-xBaTiO_3$, (x=0.00, 0.04, 0.08, 0.12) ceramics were studied by UV-visible spectroscopy. Figure 3.23 shows the graph between (hv) and $(\alpha hv)^2$ from which the optical band gap of $(1-x)(Na_{0.5}Bi_{0.5})TiO_3-xBaTiO_3$, (x=0.00, 0.04, 0.08, 0.12) ceramics are estimated. The variation in the optical band gap for different compositions is summarized in Table 3.18.

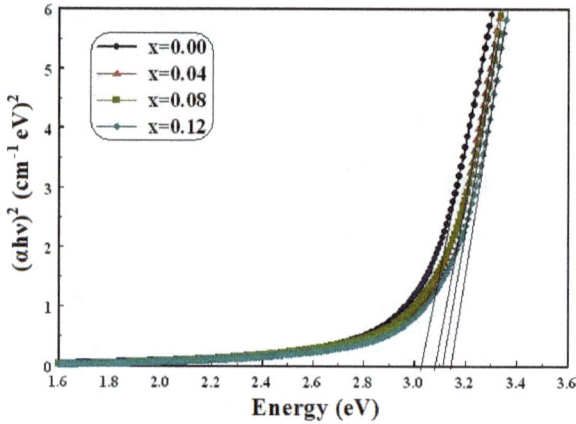

Figure 3.23 *Tauc plot for* $(1-x)(Na_{0.5}Bi_{0.5})TiO_3-xBaTiO_3$, *(x=0.00, 0.04, 0.08, 0.12) ceramics.*

Table 3.18 *Optical band gap values for (1-x)(Na$_{0.5}$Bi$_{0.5}$)TiO$_3$-xBaTiO$_3$, (x=0.00, 0.04, 0.08, 0.12) ceramics from UV-visible analysis.*

Sample concentration (x)	Band gap (eV)
x=0.00	3.027
x=0.04	3.084
x=0.08	3.110
x=0.12	3.142

3.4.4 (1-x)(K$_{0.5}$Bi$_{0.5}$)TiO$_3$-xBaTiO$_3$, (x=0.00, 0.08, 0.12)

The band gap values for (1-x)(K$_{0.5}$Bi$_{0.5}$)TiO$_3$-xBaTiO$_3$, (x=0.00, 0.08, 0.12) ceramics were determined using the Tauc plot as shown in figure 3.24. The band gap energy of the synthesized samples were obtained from the intercept of the tangent line in the plot of (αhv)^2 versus energy (hv) of the photon. Table 3.19 displays the direct band gap of (1-x)(K$_{0.5}$Bi$_{0.5}$)TiO$_3$-xBaTiO$_3$, (x=0.00, 0.08, 0.12) ceramics which varies from 3.006 eV to 3.088 eV.

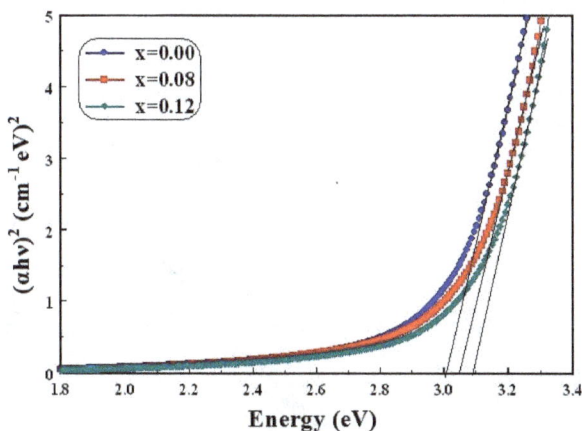

Figure 3.24 *Tauc plot for (1-x)(K$_{0.5}$Bi$_{0.5}$)TiO$_3$-xBaTiO$_3$, (x=0.00, 0.08, 0.12) ceramics.*

Table 3.19 *Optical band gap values for $(1-x)(K_{0.5}Bi_{0.5})TiO_3-xBaTiO_3$, $(x=0.00, 0.08, 0.12)$ ceramics from UV-visible analysis.*

Sample concentration (x)	Band gap (eV)
x=0.00	3.006
x=0.08	3.050
x=0.12	3.088

3.4.5 $(1-x)Ba(Zr_{0.2}Ti_{0.8})O_3-x(Ba_{0.7}Ca_{0.3})TiO_3$, $(x=0.4, 0.5, 0.6)$

The optical band gap energy (E_g) of $(1-x)Ba(Zr_{0.2}Ti_{0.8})O_3-x(Ba_{0.7}Ca_{0.3})TiO_3$, $(x=0.4, 0.5, 0.6)$ ceramics was estimated from UV-visible spectral analysis. The optical energy band gap is related to absorbance and to the photon energy, which is proposed by Wood and Tauc (Wood and Tauc, 1972). Tauc plot is drawn for the synthesized materials, by taking photon energy (hν) along the x-axis and $(αhν)^2$ along the y-axis and is shown in figure 3.25. The optical band gap values obtained from Tauc plot are tabulated in Table 3.20.

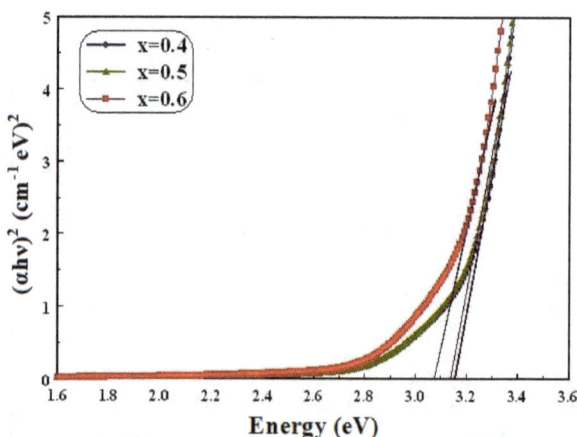

Figure 3.25 *Tauc plot for $(1-x)Ba(Zr_{0.2}Ti_{0.8})O_3-x(Ba_{0.7}Ca_{0.3})TiO_3$, $(x=0.4, 0.5, 0.6)$ ceramics.*

Table 3.20 *Optical band gap values for (1-x)Ba(Zr$_{0.2}$Ti$_{0.8}$)O$_3$-x(Ba$_{0.7}$Ca$_{0.3}$)TiO$_3$, (x=0.4, 0.5, 0.6) ceramics from UV-visible analysis.*

Sample concentration (x)	Band gap (eV)
x=0.4	3.154
x=0.5	3.137
x=0.6	3.074

3.5 Dielectric characterization

3.5.1 (1-x)(Na$_{1-y}$K$_y$)NbO$_3$-xBaTiO$_3$, (x=0.1, 0.2; y=0.01, 0.05)

The temperature dependence of dielectric constant (ε) and dielectric loss (tan δ) as a function of temperature (room temperature to 500°C) of (1-x)(Na$_{1-y}$K$_y$)NbO$_3$-xBaTiO$_3$, (x=0.1, 0.2; y=0.01, 0.05) ceramics were measured at 1 kHz. Figures 3.26 (a)-(d) and 3.27 (a)-(d) illustrate the temperature dependence of the dielectric constant (ε) and dielectric loss (tan δ) for (1-x)(Na$_{1-y}$K$_y$)NbO$_3$-xBaTiO$_3$, (x=0.1, 0.2; y=0.01, 0.05) ceramics. The obtained dielectric constant (ε) and loss (tan δ) values are displayed at 1 kHz in Table 3.21.

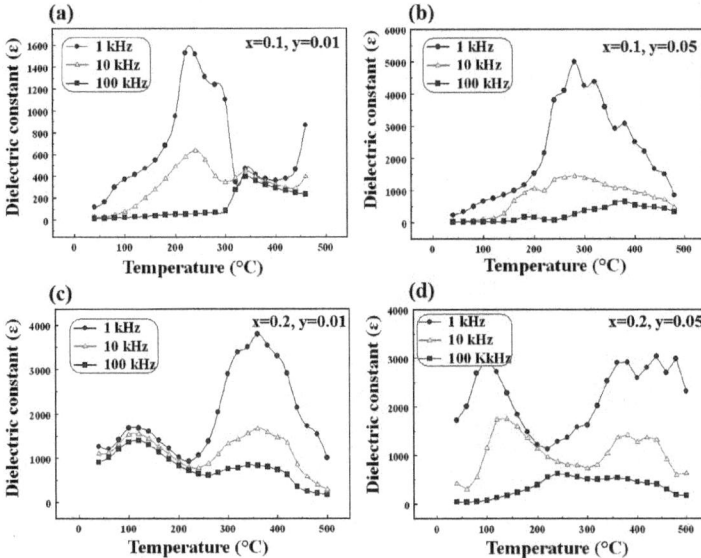

Figure 3.26 *Variation of dielectric constant (ε) as a function of temperature for (1-x)(Na$_{1-y}$K$_y$)NbO$_3$-xBaTiO$_3$, (a) x=0.1, y=0.01 (b) x=0.1, y=0.05 (c) x=0.2, y=0.01 & (d) x=0.2, y=0.05.*

Materials Research Forum LLC
doi: http://dx.doi.org/10.21741/9781945291951

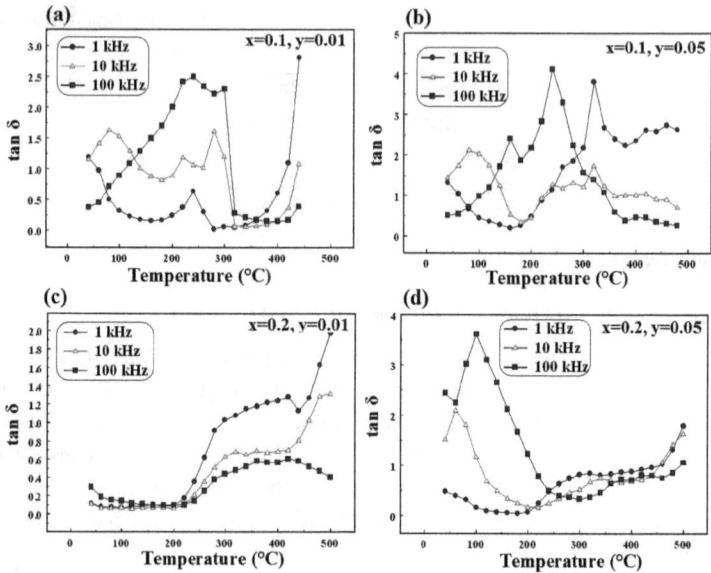

Figure 3.27 *Variation of dielectric loss (tan δ) for (1-x)(Na$_{1-y}$K$_y$)NbO$_3$-xBaTiO$_3$, (a) x=0.1, y=0.01, (b) x=0.1, y=0.05 (c) x=0.2, y=0.01 & (d) x=0.2, y=0.05 at different frequencies.*

Table 3.21 *Variation of dielectric constant (ε) and dielectric loss (tan δ) for (1-x)(Na$_{1-y}$K$_y$)NbO$_3$-xBaTiO$_3$, (x=0.1, 0.2; y=0.01, 0.05) at 1 kHz.*

Sample concentration (x, y)	Dielectric constant (ε)	Dielectric loss (tan δ)
x=0.1, y=0.01	1528	0.375
x=0.1, y=0.05	5002	1.855
x=0.2, y=0.01	3547	1.218
x=0.2, y=0.05	3031	0.960

3.5.2 (1-x)(Na$_{1-y}$K$_y$)(Nb$_{1-z}$Sb$_z$)O$_3$-xBaTiO$_3$, (x=0.1,0.2; y=0.03,0.05; z=0.05,0.1)

The dielectric constant (ε) of (1-x)(Na$_{1-y}$K$_y$)(Nb$_{1-z}$Sb$_z$)O$_3$-xBaTiO$_3$, (x=0.1, 0.2; y=0.03, 0.05; z=0.05, 0.1) ceramics as a function of temperature from room temperature to 300°C is shown in figures 3.28 (a)-(d) for the frequencies of 1 kHz, 10 kHz, 100 kHz and 1 MHz. Figures 3.29 (a)-(d) show dielectric loss (tan δ) as a function of temperature for different frequencies. A sharp dielectric peak denoting a typical ferroelectric to paraelectric phase transition is observed at 120°C, which is shown in figures 3.28 (a) and (b). Table 3.22 shows the values of dielectric constant (ε) and dielectric loss (tan δ) as a function of temperature for (1-x)(Na$_{1-y}$K$_y$)(Nb$_{1-z}$Sb$_z$)O$_3$-xBaTiO$_3$, (x=0.1, 0.2; y=0.03, 0.05; z=0.05, 0.1) ceramics at 1 kHz.

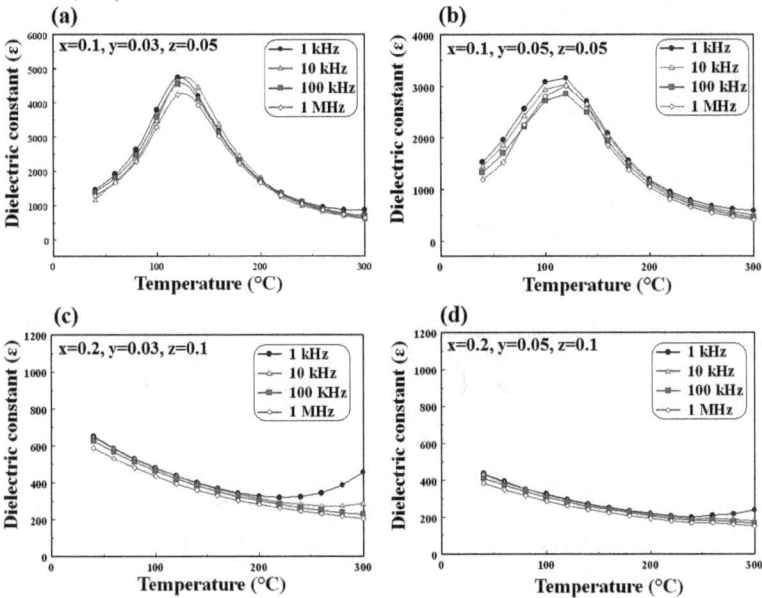

Figure 3.28 *Temperature dependence of dielectric constant for*
(1-x)(Na$_{1-y}$K$_y$)(Nb$_{1-z}$Sb$_z$)O$_3$-xBaTiO$_3$, (a) x=0.1, y=0.03, z=0.05 (b) x=0.1, y=0.05, z=0.05
(c) x=0.2, y=0.03, z=0.1 & (d) x=0.2, y=0.05, z=0.1.

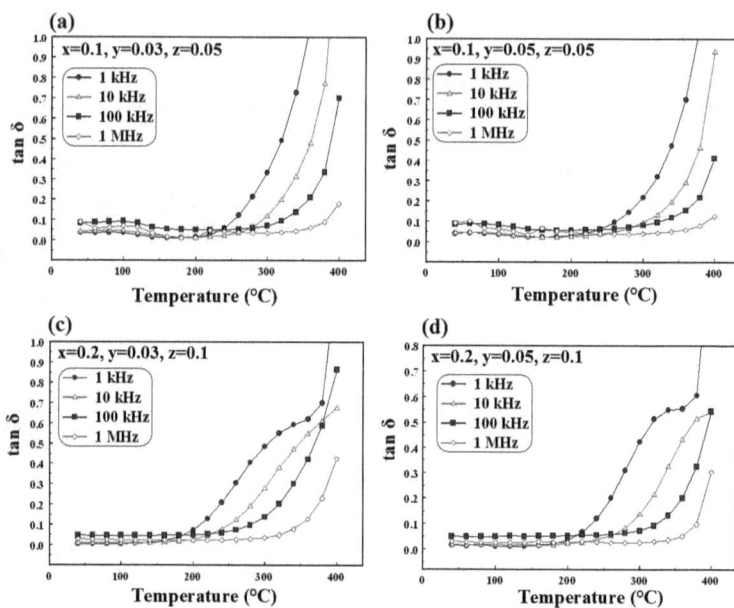

Figure 3.29 *Temperature dependence of dielectric loss (tan δ) for*
(1-x)(Na$_{1-y}$K$_y$)(Nb$_{1-z}$Sb$_z$)O$_3$-xBaTiO$_3$, (a) x=0.1, y=0.03, z=0.05 (b) x=0.1, y=0.05,
z=0.05 (c) x=0.2, y=0.03, z=0.1 & (d) x=0.2, y=0.05, z=0.1 at different frequencies

Table 3.22 *Variation of dielectric constant (ε) as a function of temperature for*
(1-x)(Na$_{1-y}$K$_y$)(Nb$_{1-z}$Sb$_z$)O$_3$-xBaTiO$_3$, (x=0.1, 0.2; y=0.03, 0.05; z=0.05, 0.1) at 1 kHz.

Sample concentration (x, y, z)	Dielectric constant (ε)	Dielectric loss (tan δ)
x=0.1, y=0.03, z=0.05	4744	0.026
x=0.1, y=0.05, z=0.05	3161	0.028
x=0.2, y=0.03, z=0.1	650	0.004
x=0.2, y=0.05, z=0.1	435	0.011

3.5.3 (1-x)(Na$_{0.5}$Bi$_{0.5}$)TiO$_3$-xBaTiO$_3$, (x=0.00, 0.04, 0.08, 0.12)

Figures 3.30 (a)-(d) and 3.31 (a)-(d) show the temperature dependent dielectric constant (ε) and dielectric loss (tan δ) for the (1-x)(Na$_{0.5}$Bi$_{0.5}$)TiO$_3$-xBaTiO$_3$, (x=0.00, 0.04, 0.08, 0.12) ceramics measured at 1 kHz, 10 kHz and 100 kHz in the temperature range from 40-400°C, respectively. The values of the dielectric constant (ε) and the dielectric loss (tan δ) at the frequency of 100 kHz for (1-x)(Na$_{0.5}$Bi$_{0.5}$)TiO$_3$-xBaTiO$_3$, (x=0.00, 0.04, 0.08, 0.12) ceramics are given in Table 3.23.

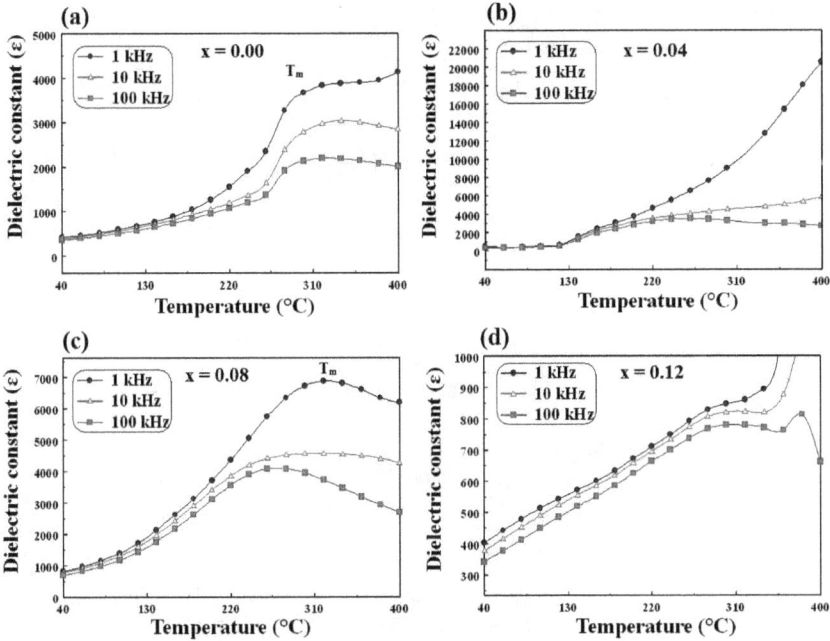

Figure 3.30 *Temperature dependence of dielectric constant (ε) for (1-x)(Na$_{0.5}$Bi$_{0.5}$)TiO$_3$-xBaTiO$_3$, (a) x=0.00 (b) x=0.04 (c) x=0.08 & (d) x=0.12 at different frequencies.*

Materials Research Forum LLC
doi: http://dx.doi.org/10.21741/9781945291951

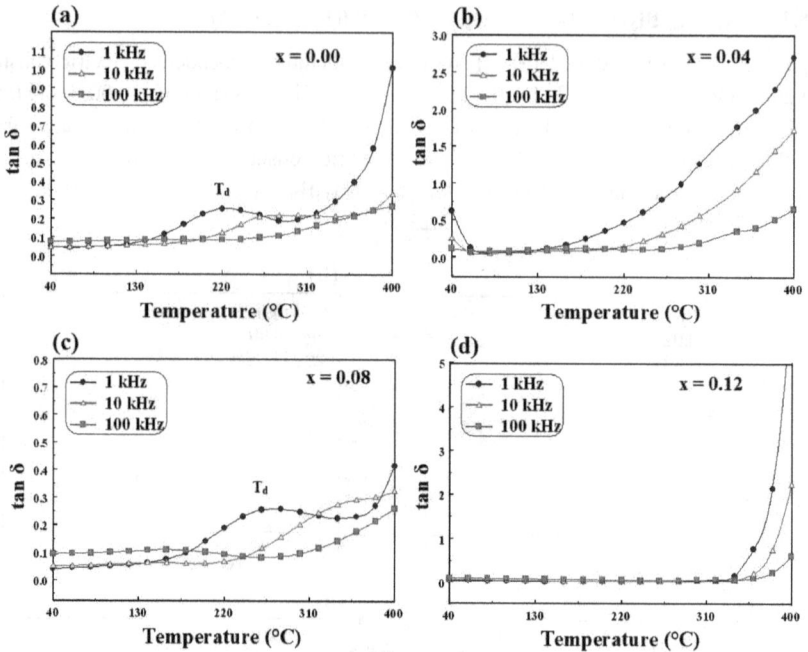

Figure 3.31 *Temperature dependence of dielectric loss (tan δ) for*
(1-x)(Na$_{0.5}$Bi$_{0.5}$)TiO$_3$-xBaTiO$_3$, **(a)** *x=0.00* **(b)** *x=0.04* **(c)** *x=0.08 &* **(d)** *x=0.12 at*
different frequencies.

Table 3.23 *Variation of dielectric constant (ε) and dielectric loss (tan δ) for*
(1-x)(Na$_{0.5}$Bi$_{0.5}$)TiO$_3$-xBaTiO$_3$, (x=0.00, 0.04, 0.08, 0.12) at 100 kHz.

Sample concentration (x)	Dielectric constant (ε)	Dielectric loss (tan δ)
x=0.00	2199	0.163
x=0.04	3500	0.111
x=0.08	4070	0.082
x=0.12	779	0.057

3.5.4 (1-x)(K₀.₅Bi₀.₅)TiO₃-xBaTiO₃, (x=0.00, 0.08, 0.12)

Figure 3.32 (a) shows the variation of dielectric constant (ε) of $(1\text{-}x)(K_{0.5}Bi_{0.5})TiO_3$-$xBaTiO_3$, (x=0.00, 0.08, 0.12) ceramics with respect to the frequency ranging from 10 Hz to 1 MHz measured at room temperature. Figure 3.32 (b) shows the variation of dielectric loss (tan δ) as a function of frequency for $(1\text{-}x)(K_{0.5}Bi_{0.5})TiO_3$-$xBaTiO_3$, (x=0.00, 0.08, 0.12) ceramics. The values of dielectric constant (ε) and dielectric loss (tan δ) for undoped $K_{0.5}Bi_{0.5}TiO_3$ are found to be 511 and 0.51 respectively at 10 kHz. Table 3.24 shows values of dielectric constant (ε) and dielectric loss (tan δ) for all the samples of $(1\text{-}x)(K_{0.5}Bi_{0.5})TiO_3$-$xBaTiO_3$, (x=0.00, 0.08, 0.12) ceramics at 10 kHz.

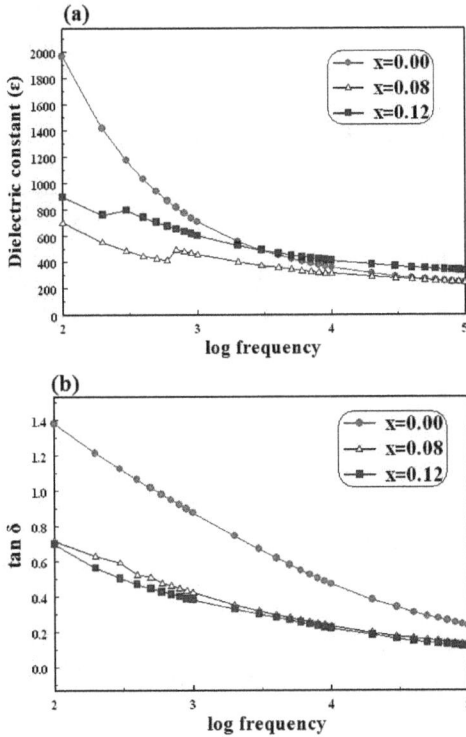

Figure 3.32 *Frequency dependence of dielectric properties of $(1\text{-}x)(K_{0.5}Bi_{0.5})TiO_3$-$xBaTiO_3$, (x=0.00, 0.08, 0.12)* ***(a)*** *dielectric constant &* ***(b)*** *dielectric loss at a frequency of 10 kHz.*

Materials Research Forum LLC
doi: http://dx.doi.org/10.21741/9781945291951

Table 3.24 *Variation of dielectric constant (ε) and dielectric loss (tan δ) for*
(1-x)(K$_{0.5}$Bi$_{0.5}$)TiO$_3$-xBaTiO$_3$, (x=0.00, 0.08, 0.12) at 10 kHz.

Sample concentration (x)	Dielectric constant (ε)	Dielectric loss (tan δ)
x=0.00	511	0.512
x=0.08	418	0.260
x=0.12	529	0.301

3.5.5 (1-x)Ba(Zr$_{0.2}$Ti$_{0.8}$)O$_3$-x(Ba$_{0.7}$Ca$_{0.3}$)TiO$_3$, (x=0.4, 0.5, 0.6)

Figure 3.33 shows the temperature dependence of dielectric constant (ε) of the lead-free (1-x)Ba(Zr$_{0.2}$Ti$_{0.8}$)O$_3$-x(Ba$_{0.7}$Ca$_{0.3}$)TiO$_3$, (x=0.4, 0.5, 0.6) ceramics measured at 10 kHz. The dielectric constant (ε) was measured in the temperature range from 40°C to 200°C. The values of the dielectric constant (ε) at 10 kHz for (1-x)Ba(Zr$_{0.2}$Ti$_{0.8}$)O$_3$-x(Ba$_{0.7}$Ca$_{0.3}$)TiO$_3$, (x=0.4, 0.5, 0.6) ceramics are given in Table 3.25.

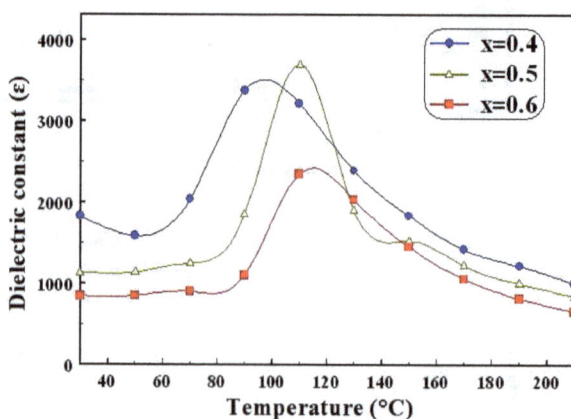

Figure 3.33 *Temperature dependence of dielectric constant (ε) for (1-x)Ba(Zr$_{0.2}$Ti$_{0.8}$)O$_3$-x(Ba$_{0.7}$Ca$_{0.3}$)TiO$_3$, (x=0.4, 0.5, 0.6) at a frequency of 10 kHz.*

Lead-free Piezo-Ceramic Solid Solutions, R. Saravanan   Materials Research Forum LLC
Materials Research Foundations **41** (2018)   doi: http://dx.doi.org/10.21741/9781945291951

Table 3.25 *Variation of dielectric constant (ε) and dielectric loss (tan δ) for*
(1-x)Ba(Zr$_{0.2}$Ti$_{0.8}$)O$_3$-x(Ba$_{0.7}$Ca$_{0.3}$)TiO$_3$, (x=0.4, 0.5, 0.6) at a frequency of 10 kHz.

Sample concentration (x)	Dielectric constant (ε)
x=0.4	3374
x=0.5	3702
x=0.6	2151

3.6 Ferroelectric and piezoelectric measurements

3.6.1 (1-x)(Na$_{1-y}$K$_y$)NbO$_3$-xBaTiO$_3$, (x=0.1, 0.2; y=0.01, 0.05)

Figures 3.34 (a)-(d) show the ferroelectric hysteresis loops of (1-x)(Na$_{1-y}$K$_y$)NbO$_3$-xBaTiO$_3$, (x=0.1, 0.2; y=0.01, 0.05) ceramic samples. The values of maximum polarization (P$_m$), remnant polarization (P$_r$) and coercive field (E$_C$) determined from the hysteresis loops are summarized in Table 3.26. The piezoelectric constant (d$_{33}$) of (1-x)(Na$_{1-y}$K$_y$)NbO$_3$-xBaTiO$_3$, (x=0.1, 0.2; y=0.01, 0.05) ceramic samples are also given in Table 3.26.

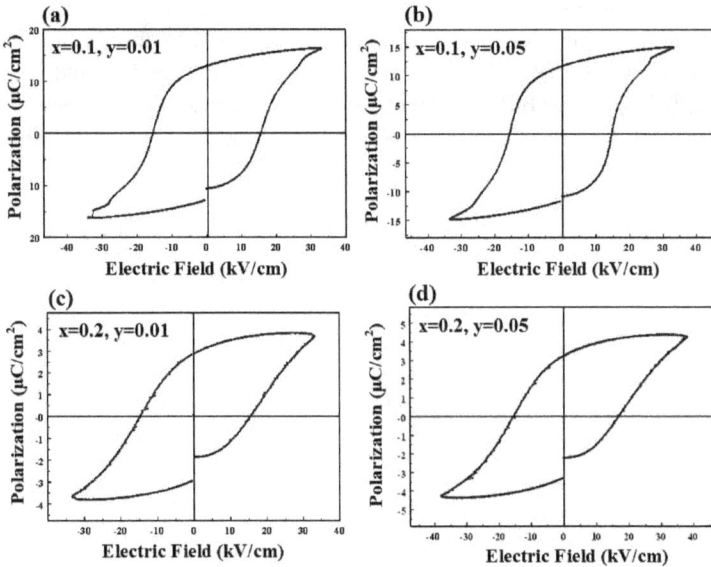

Figure 3.34 *Hysteresis loops for (1-x)(Na$_{1-y}$K$_y$)NbO$_3$-xBaTiO$_3$ ceramics (a) x=0.1, y=0.01 (b) x=0.1, y=0.05 (c) x=0.2, y=0.01 & (d) x=0.2, y=0.05.*

Table 3.26 *Ferroelectric and piezoelectric properties of*
(1-x)(Na$_{1-y}$K$_y$)NbO$_3$-xBaTiO$_3$, (x=0.1, 0.2; y=0.01, 0.05) ceramics.

Sample concentration (x, y)	P$_m$ (μC/cm^2)	E$_C$ (kV/cm)	P$_r$ (μC/cm^2)	d$_{33}$ (pC/N)
x=0.1, y=0.01	16.32	15.73	12.88	120
x=0.1, y=0.05	14.91	14.86	11.56	104
x=0.2, y=0.01	3.82	15.78	2.86	38
x=0.2, y=0.05	4.38	16.91	3.22	34

P$_m$ - maximum polarization
E$_C$ - coercive field
P$_r$ - remnant polarization
d$_{33}$ - piezoelectric constant

3.6.2 (1-x)(Na$_{1-y}$K$_y$)(Nb$_{1-z}$Sb$_z$)O$_3$-xBaTiO$_3$, (x=0.1,0.2; y=0.03,0.05; z=0.05,0.1)

The composition dependence of polarization versus electric field hysteresis loops for (1-x)(Na$_{1-y}$K$_y$)(Nb$_{1-z}$Sb$_z$)O$_3$-xBaTiO$_3$, (x=0.1, 0.2; y=0.03, 0.05; z=0.05, 0.1) ceramics were measured at room temperature and are shown in figures 3.35 (a)-(d). The values of maximum polarization (P$_m$), remnant polarization (P$_r$) and coercive field (E$_C$) of (1-x)(Na$_{1-y}$K$_y$)(Nb$_{1-z}$Sb$_z$)O$_3$-xBaTiO$_3$ were obtained from hysteresis loops and given in Table 3.27. The piezoelectric constant (d$_{33}$) of (1-x)(Na$_{1-y}$K$_y$)(Nb$_{1-z}$Sb$_z$)O$_3$-xBaTiO$_3$, (x=0.1, 0.2; y=0.03, 0.05; z=0.05, 0.1) ceramics are also given in Table 3.27.

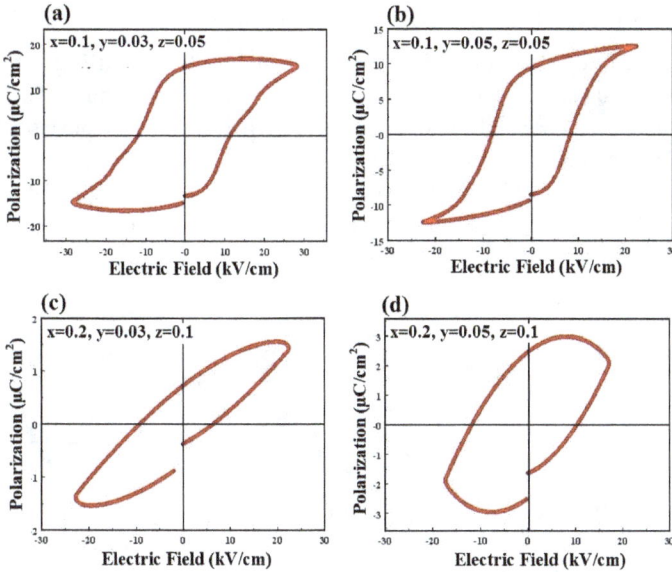

Figure 3.35 *Hysteresis loops for $(1-x)(Na_{1-y}K_y)(Nb_{1-z}Sb_z)O_3$-$xBaTiO_3$ ceramics* ***(a)*** *x=0.1, y=0.03, z=0.05* ***(b)*** *x=0.1, y=0.05, z=0.05* ***(c)*** *x=0.2, y=0.03, z=0.1 &* ***(d)*** *x=0.2, y=0.05, z=0.1.*

Table 3.27 *Ferroelectric and piezoelectric properties of $(1-x)(Na_{1-y}K_y)(Nb_{1-z}Sb_z)O_3$-$xBaTiO_3$, (x=0.1, 0.2; y=0.03, 0.05; z=0.05, 0.1) ceramics.*

Sample concentration (x, y, z)	P_m ($\mu C/cm^2$)	E_C (kV/cm)	P_r ($\mu C/cm^2$)	d_{33} (pC/N)
x=0.1, y=0.03, z=0.05	16.73	11.47	14.83	110
x=0.1, y=0.05, z=0.05	12.49	8.29	9.28	94
x=0.2, y=0.03, z=0.1	3.81	15.72	2.86	54
x=0.2, y=0.05, z=0.1	4.43	3.35	3.74	35

P_m - maximum polarization
E_C - coercive field
P_r - remnant polarization
d_{33} - piezoelectric constant

Materials Research Forum LLC
doi: http://dx.doi.org/10.21741/9781945291951

3.6.3 (1-x)(Na$_{0.5}$Bi$_{0.5}$)TiO$_3$-xBaTiO$_3$, (x=0.00, 0.04, 0.08, 0.12)

Figures 3.36 (a)-(d) show the P-E hysteresis loops of (1-x)(Na$_{0.5}$Bi$_{0.5}$)TiO$_3$-xBaTiO$_3$, (x=0.00, 0.04, 0.08, 0.12) ceramics measured at room temperature. The saturated hysteresis loop was observed for x=0.00. The coercive field (E$_C$), remnant polarization (P$_r$) and the maximum polarization (P$_m$) are extracted from figures 3.36 (a)-(d) and are listed in Table 3.28. The piezoelectric coefficient (d$_{33}$) of (1-x)(Na$_{0.5}$Bi$_{0.5}$)TiO$_3$-xBaTiO$_3$, (x=0.00, 0.04, 0.08, 0.12) ceramics were measured using a d$_{33}$ meter at room temperature and the d$_{33}$ values are also given in Table 3.28.

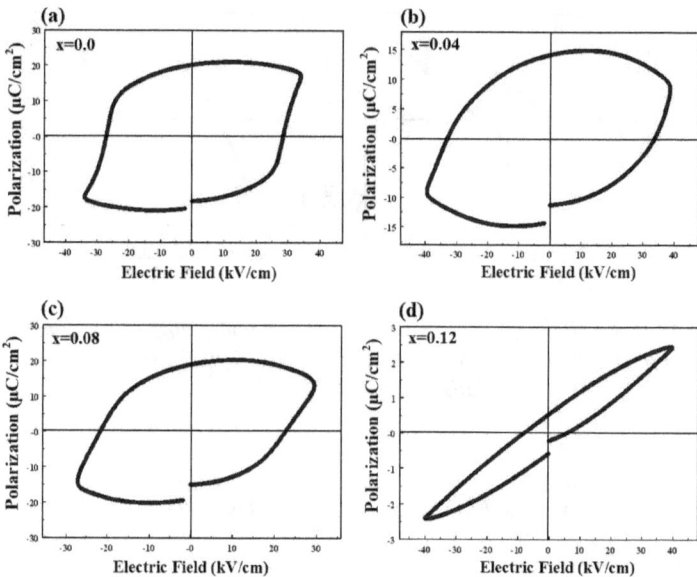

Figure 3.36 *Ferroelectric P-E hysteresis loops of (1-x)(Na$_{0.5}$Bi$_{0.5}$)TiO$_3$-xBaTiO$_3$ ceramics at room temperature,* **(a)** *x=0.00* **(b)** *x=0.04* **(c)** *x=0.08* & **(d)** *x=0.12.*

Table 3.28 *Ferroelectric and piezoelectric properties of*
$(1-x)(Na_{0.5}Bi_{0.5})TiO_3-xBaTiO_3$, $(x=0.00, 0.04, 0.08, 0.12)$ ceramics.

Sample concentration (x)	P_m $(\mu C/cm^2)$	E_C (kV/cm)	P_r $(\mu C/cm^2)$	d_{33} (pC/N)
x=0.00	21.45	28.77	20.60	63
x=0.04	14.84	33.81	14.09	78
x=0.08	20.24	22.68	18.92	122
x=0.12	2.43	5.67	0.53	58

P_m - maximum polarization
E_C - coercive field
P_r - remnant polarization
d_{33} - piezoelectric constant

3.6.4 $(1-x)(K_{0.5}Bi_{0.5})TiO_3-xBaTiO_3$, $(x=0.00, 0.08, 0.12)$

Polarization-electric hysteresis loops of the $(1-x)(K_{0.5}Bi_{0.5})TiO_3-xBaTiO_3$, $(x=0.00, 0.08, 0.12)$ ceramics were measured at room temperature as shown in figures 3.37 (a)-(c). The values of coercive field (E_C), remnant polarization (P_r) and maximum polarization (P_m) are given in Table 3.29.

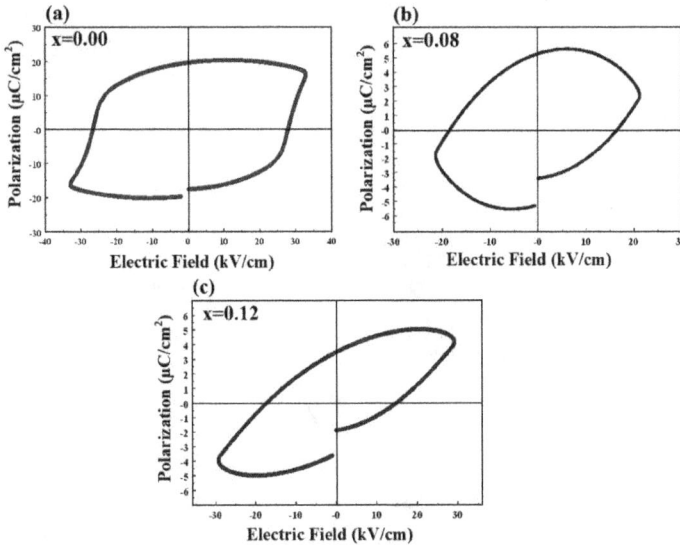

Figure 3.37 *P-E hysteresis loops of $(1-x)(K_{0.5}Bi_{0.5})TiO_3-xBaTiO_3$, **(a)** x=0.00*
***(b)** x=0.08 **(c)** x=0.12 ceramics measured at room temperatures.*

Table 3.29 *Ferroelectric and piezoelectric properties of*
(1-x)(K$_{0.5}$Bi$_{0.5}$)TiO$_3$-xBaTiO$_3$, (x=0.00, 0.08, 0.12) ceramics.

Sample concentration (x)	P$_m$ (μC/cm^2)	P$_r$ (μC/cm^2)	E$_C$ (kV/cm)
x=0.00	20.24	19.47	5.22
x=0.08	5.58	5.24	16.64
x=0.12	5.01	3.47	15.03

P$_m$ - maximum polarization
P$_r$ - remnant polarization
E$_C$ - coercive field

3.6.5 (1-x)Ba(Zr$_{0.2}$Ti$_{0.8}$)O$_3$-x(Ba$_{0.7}$Ca$_{0.3}$)TiO$_3$, (x=0.4, 0.5, 0.6)

Figures 3.38 (a)-(c) show the P-E hysteresis loops of (1-x)Ba(Zr$_{0.2}$Ti$_{0.8}$)O$_3$-x(Ba$_{0.7}$Ca$_{0.3}$)TiO$_3$, (x=0.4, 0.5, 0.6) ceramics measured at room temperature. The maximum polarization (P$_m$), remnant polarization (P$_r$) and the coercive field (E$_C$) of the samples are given in Table 3.30. Figure 3.38 (b) shows soft P-E hysteresis loop (x=0.5) with a maximum polarization (P$_m$) of 15.12 μC/cm^2 and a coercive filed (E$_C$) of 3.62 kV/cm. 0.5BZT-0.5BCT composition exhibits the maximum polarization at room temperature. The values of piezoelectric charge coefficient (d$_{33}$) of (1-x)Ba(Zr$_{0.2}$Ti$_{0.8}$)O$_3$-x(Ba$_{0.7}$Ca$_{0.3}$)TiO$_3$, (x=0.4, 0.5, 0.6) ceramics are given in Table 3.30.

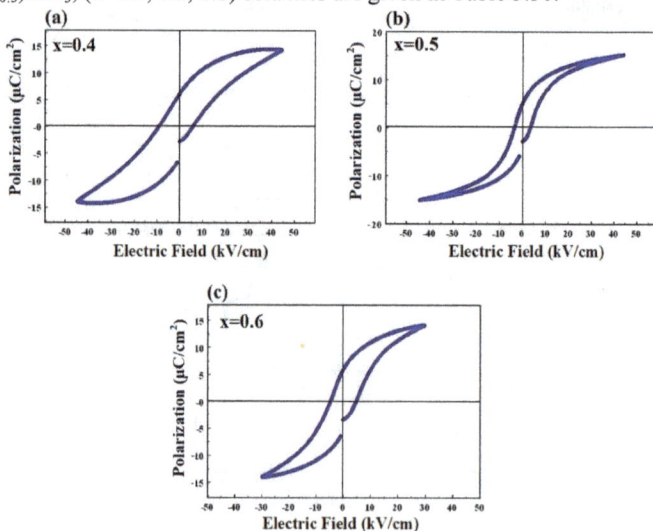

Figure 3.38 *P-E hysteresis loops of (1-x)Ba(Zr$_{0.2}$Ti$_{0.8}$)O$_3$-x(Ba$_{0.7}$Ca$_{0.3}$)TiO$_3$, **(a)** x=0.4*
***(b)** x=0.5 **(c)** x=0.6 ceramics measured at room temperatures.*

Table 3.30 *Ferroelectric and piezoelectric properties of*
(1-x)Ba(Zr$_{0.2}$Ti$_{0.8}$)O$_3$-x(Ba$_{0.7}$Ca$_{0.3}$)TiO$_3$, (x=0.4, 0.5, 0.6) ceramics.

Sample concentration (x)	P$_m$ (μC/cm^2)	P$_r$ (μC/cm^2)	E$_c$ (kV/cm)	d$_{33}$ (pC/N)
x=0.4	14.37	5.98	6.02	215
x=0.5	15.12	4.90	3.62	276
x=0.6	14.06	5.76	4.88	176

P$_m$ - maximum polarization
P$_r$ - remnant polarization
E$_C$ - coercive field
d$_{33}$ - piezoelectric constant

3.7 Electron density studies

The electron distribution study is an important part in materials characterization. Electronic distribution around the atoms in the unit cell plays a vital role in the nature of materials and it controls the physical properties of the materials. Maximum entropy method (MEM) (Collins, 1982) is an important and accurate technique to deal the electron density distribution in the unit cell because of their probabilistic approach. The structure factors extracted from Rietveld method (Rietveld, 1969) were used to reconstruct the electron density distribution in the unit cell. Also, it only needs a minimum amount of information from the observed XRD spectra and it yields least biased information. This method is packaged by the software PRactice Iterative MEM Analyses (PRIMA) (Momma and Izumi, 2011). In this book, the charge density results are visualized by the visualization software VESTA (Visualization for Electronic and STructural Analysis) (Momma and Izumi, 2011).

3.7.1 (1-x)(Na$_{1-y}$K$_y$)NbO$_3$-xBaTiO$_3$, (x=0.1, 0.2; y=0.01, 0.05)

Three dimensional charge density distributions in the unit cell of lead-free (1-x)(Na$_{1-y}$K$_y$)NbO$_3$-xBaTiO$_3$, (x=0.1, 0.2; y=0.01, 0.05) are constructed with similar isosurface levels of 1 e/Å^3 and are presented in figures 3.39 (a)-(d) with (002) plane shaded. The two dimensional electron density distributions on (001) plane are shown in figures 3.40 (a)-(d). Figures 3.41 (a)-(d) show the two dimensional charge density distributions on (002) plane, with the enlarged view of electron density distribution around Nb and O atoms. These results are quantitatively analyzed by plotting one dimensional charge density profiles along Na-O and Nb-O, which are shown in figures 3.42 (a) and (b). The mid bond electron density values are tabulated in Table 3.31.

Materials Research Forum LLC
doi: http://dx.doi.org/10.21741/9781945291951

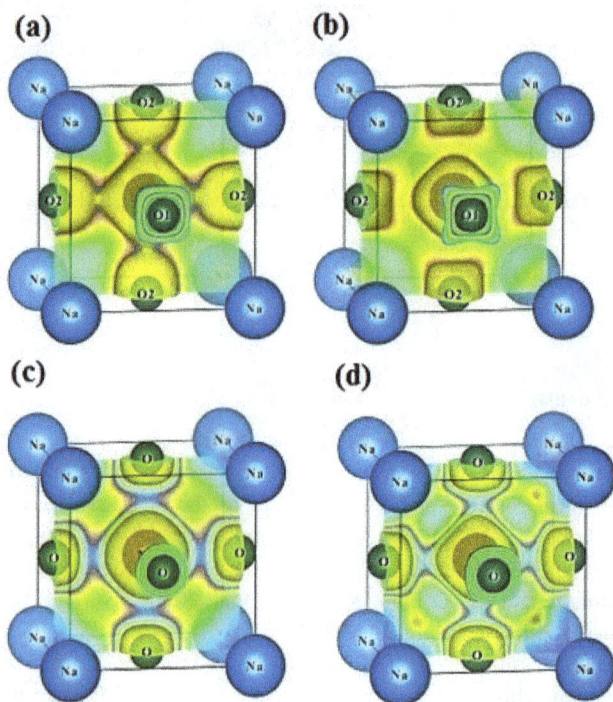

Figure 3.39 Three dimensional charge density isosurfaces for (1-x)(Na$_{1-y}$K$_y$)NbO$_3$-xBaTiO$_3$ with (002) plane shaded for **(a)** x=0.1, y=0.01 (Tetragonal) **(b)** x=0.1, y=0.05 (Tetragonal) **(c)** x=0.2, y=0.01 (cubic) & **(d)** x=0.2, y=0.05 (cubic)

Materials Research Forum LLC
doi: http://dx.doi.org/10.21741/9781945291951

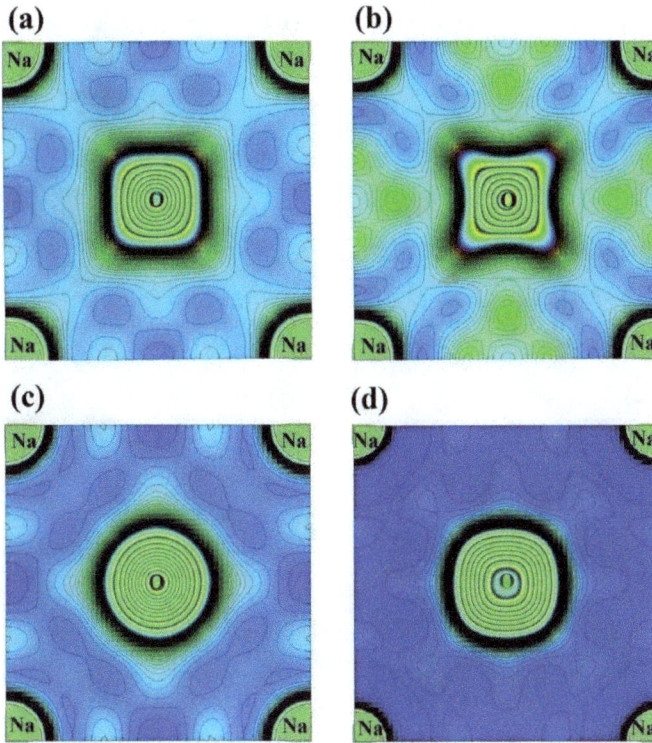

Figure 3.40 Two dimensional charge density distribution on (001) plane for
$(1-x)(Na_{1-y}K_y)NbO_3$-$xBaTiO_3$, *(a)* $x=0.1$, $y=0.01$ *(b)* $x=0.1$, $y=0.05$ *(c)* $x=0.2$, $y=0.01$
& *(d)* $x=0.2$, $y=0.05$

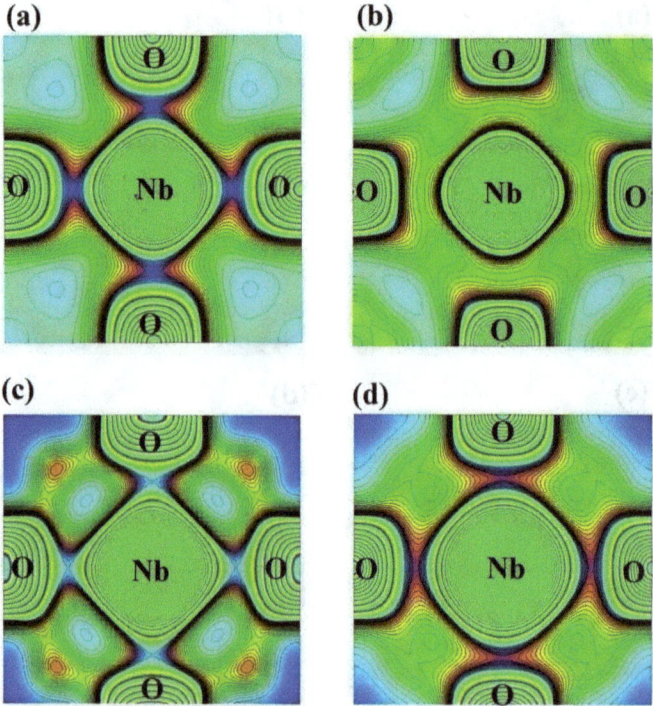

Figure 3.41 *Two dimensional charge density distribution on (002) plane for*
(1-x)(Na$_{1-y}$K$_y$)NbO$_3$-xBaTiO$_3$, ***(a)*** *x=0.1, y=0.01* ***(b)*** *x=0.1, y=0.05* ***(c)*** *x=0.2, y=0.01*
*& **(d)** x=0.2, y=0.05*

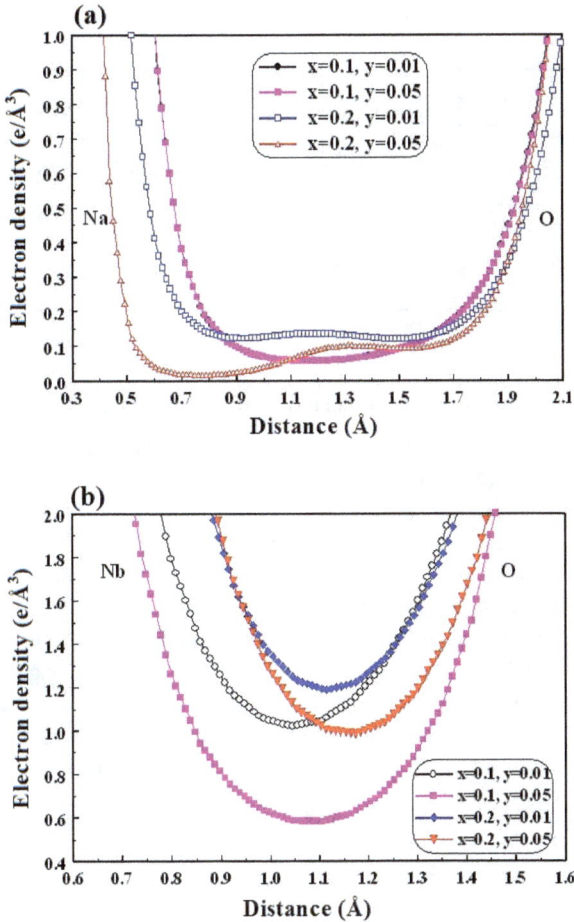

Figure 3.42 (a) *One dimensional electron density profiles between Na and O atoms*
(b) *One dimensional electron density profiles between Nb and O atoms.*

Table 3.31 *Bond lengths and mid bond electron densities for Na-O and Nb-O bond of*
(1-x)(Na$_{1-y}$K$_y$)NbO$_3$-xBaTiO$_3$, (x=0.1, 0.2; y=0.01, 0.05).

Sample concentration (x, y)	Bonding			
	Na-O		Nb-O	
	Bond length (Å)	Mid bond electron density (e/Å^3)	Bond length (Å)	Mid bond electron density (e/Å^3)
x=0.1, y=0.01	2.7873	0.0697	1.9632	1.0558
x=0.1, y=0.05	2.7901	0.0715	1.9633	0.6317
x=0.2, y=0.01	2.7974	0.1223	1.9787	1.3644
x=0.2, y=0.05	2.8011	0.0963	1.9813	1.3406

3.7.2   (1-x)(Na$_{1-y}$K$_y$)(Nb$_{1-z}$Sb$_z$)O$_3$-xBaTiO$_3$, (x=0.1,0.2; y=0.03,0.05; z=0.05,0.1)

Figures 3.43 (a)-(d) show the three dimensional electron density distribution in the unit cell of (1-x)(Na$_{1-y}$K$_y$)(Nb$_{1-z}$Sb$_z$)O$_3$-xBaTiO$_3$, (x=0.1, 0.2; y=0.03, 0.05; z=0.05, 0.1) ceramics with the isosurface level of 1.5 e/Å^3. Figures 3.43 (a)-(d) show the electron clouds around the Na, Nb and O atoms. Figures 3.44 (a)-(d) show the two dimensional charge density distributions and contour lines around Na-O bond and figures 3.45 (a)-(d) show the two dimensional charge density distributions and contour lines around Nb-O bond. Figures 3.46 (a) and (b) show the one dimensional charge density profiles along Na-O bonding direction (Figure 3.46 (a)) and along the Nb-O bonding direction (Figure 3.46 (b)), respectively. The one dimensional profiles are used to analyze the bonding nature between Na-O and Nb-O bonds of (1-x)(Na$_{1-y}$K$_y$)(Nb$_{1-z}$Sb$_z$)O$_3$-xBaTiO$_3$ ceramics. Table 3.32 represents the bond length and mid bond electron density values.

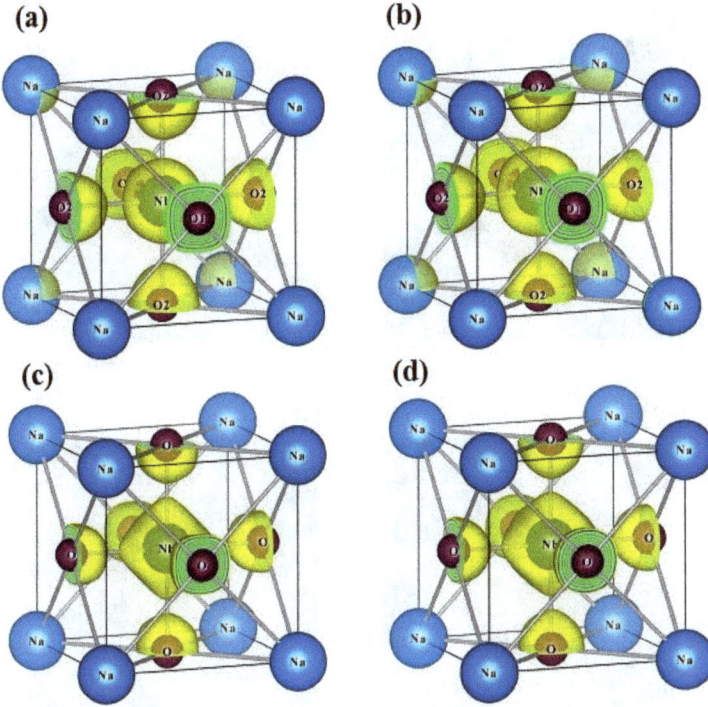

Figure 3.43 *Three dimensional charge density isosurfaces for*
(1-x)(Na$_{1-y}$K$_y$)(Nb$_{1-z}$Sb$_z$)O$_3$-xBaTiO$_3$ for **(a)** *x=0.1, y=0.03, z=0.05 (tetragonal)* **(b)** *x=0.1,*
y=0.05, z=0.05 (tetragonal) **(c)** *x=0.2, y=0.03, z=0.1 (cubic) &*
(d) *x=0.2, y=0.05, z=0.1 (cubic).*

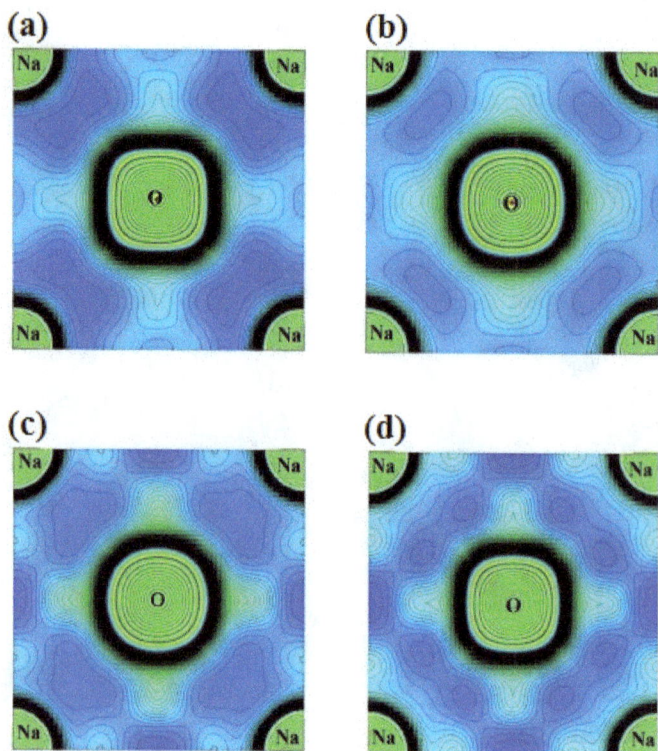

Figure 3.44 *Two dimensional charge density distribution on (001) plane for*
(1-x)(Na$_{1-y}$K$_y$)(Nb$_{1-z}$Sb$_z$)O$_3$-xBaTiO$_3$ (a) x=0.1, y=0.03, z=0.05 (b) x=0.1, y=0.05, z=0.05
c) x=0.2, y=0.03, z=0.1 & (d) x=0.2, y=0.05, z=0.1

Materials Research Forum LLC
doi: http://dx.doi.org/10.21741/9781945291951

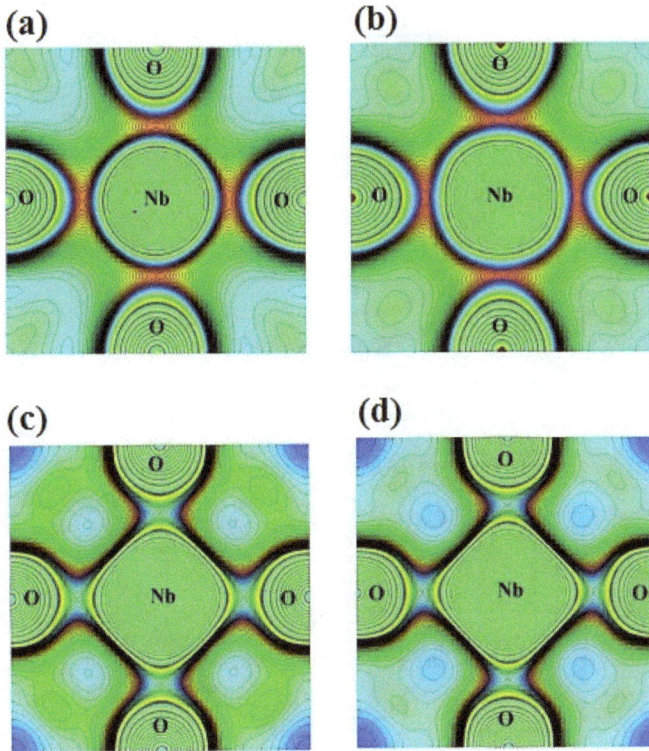

Figure 3.45 *Two dimensional charge density distribution on (002) plane for*
*(1-x)(Na$_{1-y}$K$_y$)(Nb$_{1-z}$Sb$_z$)O$_3$-xBaTiO$_3$, **(a)** x=0.1, y=0.03, z=0.05 **(b)** x=0.1, y=0.05, z=0.05*
***(c)** x=0.2, y=0.03, z=0.1 & **(d)** x=0.2, y=0.05, z=0.1.*

Materials Research Forum LLC
doi: http://dx.doi.org/10.21741/9781945291951

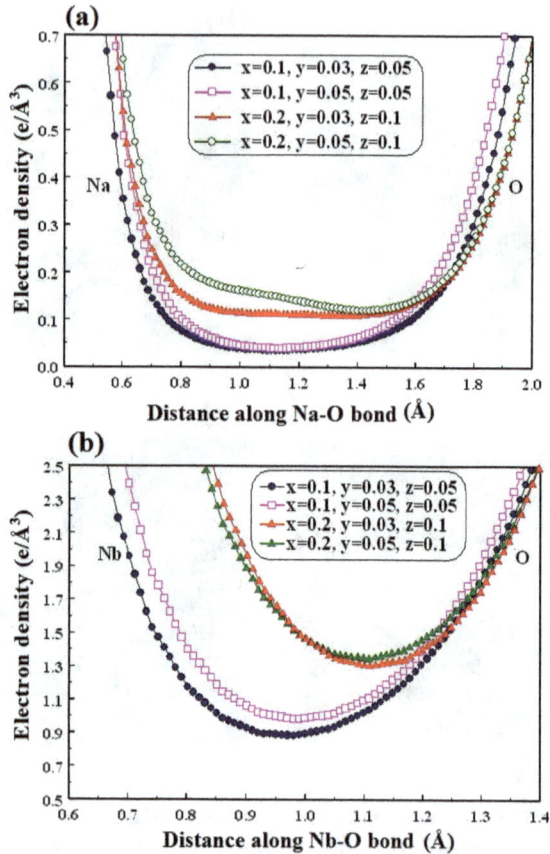

Figure 3.46 (a) *One dimensional electron density profiels along [Na-O] bonding direction **(b)** One dimensional electron density profiles along [Nb-O] bonding direction.*

Table 3.32 *Bond lengths and mid bond electron densities for Na-O and Nb-O bond for*
(1-x)(Na$_{1-y}$K$_y$)(Nb$_{1-z}$Sb$_z$)O$_3$-xBaTiO$_3$, (x=0.1, 0.2; y=0.03, 0.05; z=0.05, 0.1)

Sample concentration (x, y, z)	Bonding			
	Na-O		Nb-O	
	Bond length (Å)	Mid bond electron density (e/Å^3)	Bond length (Å)	Mid bond electron density (e/Å^3)
x=0.1, y=0.03, z=0.05	2.7890	0.0572	1.9733	0.9775
x=0.1, y=0.05, z=0.05	2.7819	0.0604	1.9707	1.0133
x=0.2, y=0.03, z=0.1	2.7852	0.1092	1.9694	1.4875
x=0.2, y=0.05, z=0.1	2.7889	0.1198	1.9721	1.4886

3.7.3 (1-x)(Na$_{0.5}$Bi$_{0.5}$)TiO$_3$-xBaTiO$_3$, (x=0.00, 0.04, 0.08, 0.12)

Figures 3.47 (a)-(d) show the three dimensional charge density distributions in the unit cell with similar isosurface level of 1 e/Å^3. Figures 3.47 (a)-(d) confirms that the (1-x)(Na$_{0.5}$Bi$_{0.5}$)TiO$_3$-xBaTiO$_3$ ceramics were crystallized with rhombohedral and tetragonal structures. The two dimensional electron density distributions on (1$\bar{2}$0) and (001) planes are drawn in the contour range of 0 to 1 e/Å^3 with contour interval of 0.05 e/Å^3 and are presented in figures 3.48 (a)-(d) for (1$\bar{2}$0) plane and 3.49 (a)-(d) for (001) plane respectively. Figures 3.50 (a) and (b) show the one dimensional charge density profiles along Na-O and Ti-O bonds respectively. The bond lengths and mid bond electron density values between the Na-O and Ti-O bonds are presented in Table 3.33.

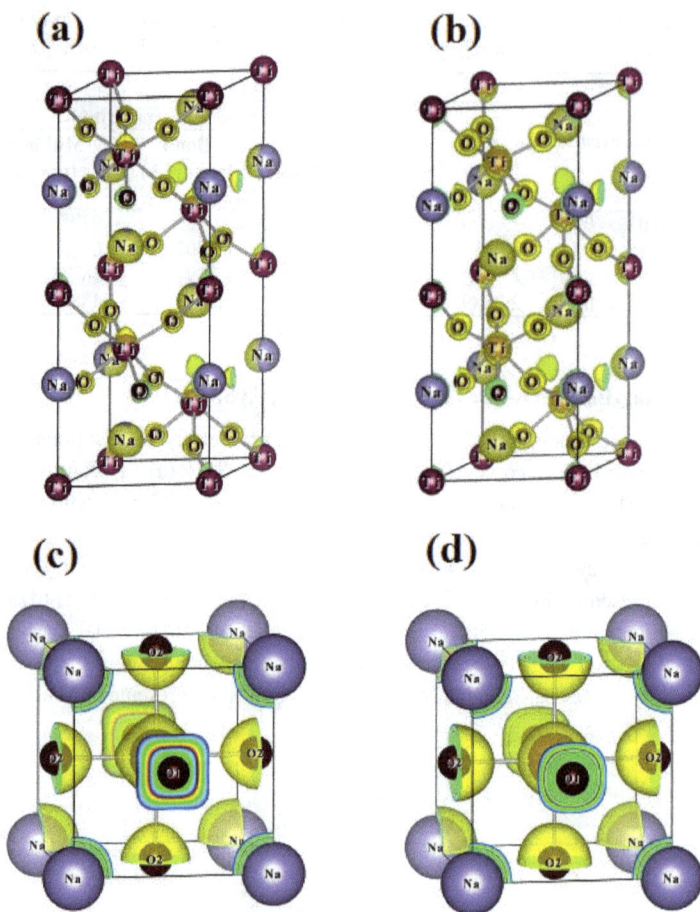

Figure 3.47 *Three dimensional charge density distribution for $(1-x)(Na_{0.5}Bi_{0.5})TiO_3$-$xBaTiO_3$,* **(a)** *x=0.00 (rhombohedral)* **(b)** *x=0.04 (rhombohedral)* **(c)** *x=0.08 (tetragonal) & **(d)** x=0.12 (tetragonal).*

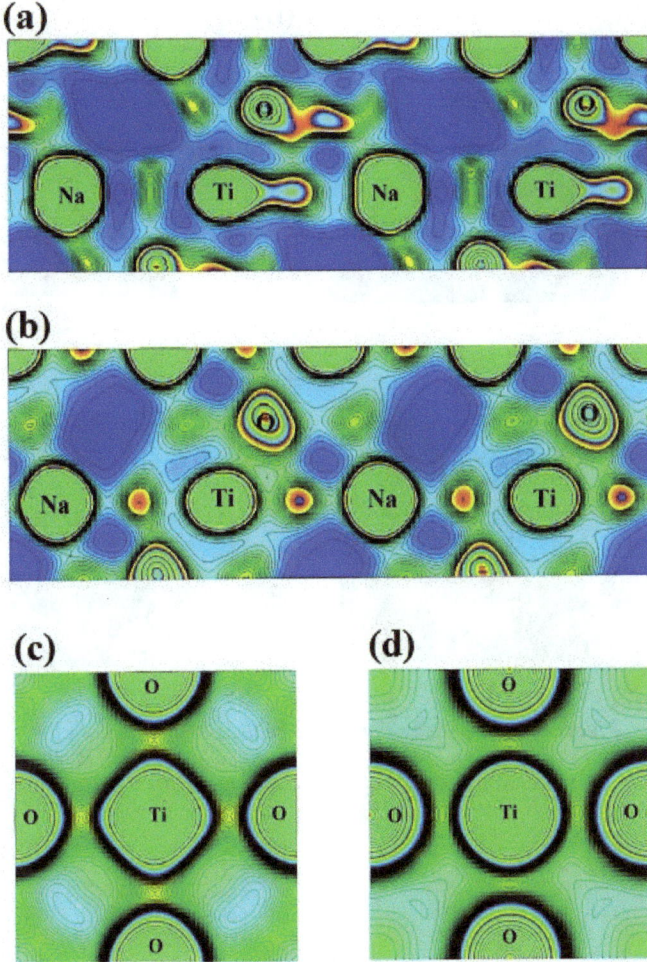

Figure 3.48 *Two dimensional charge density distribution on (001) plane for*
*(1-x)(Na$_{0.5}$Bi$_{0.5}$)TiO$_3$-xBaTiO$_3$, **(a)** x=0.00 **(b)** x=0.04 **(c)** x=0.08 & **(d)** x=0.12.*

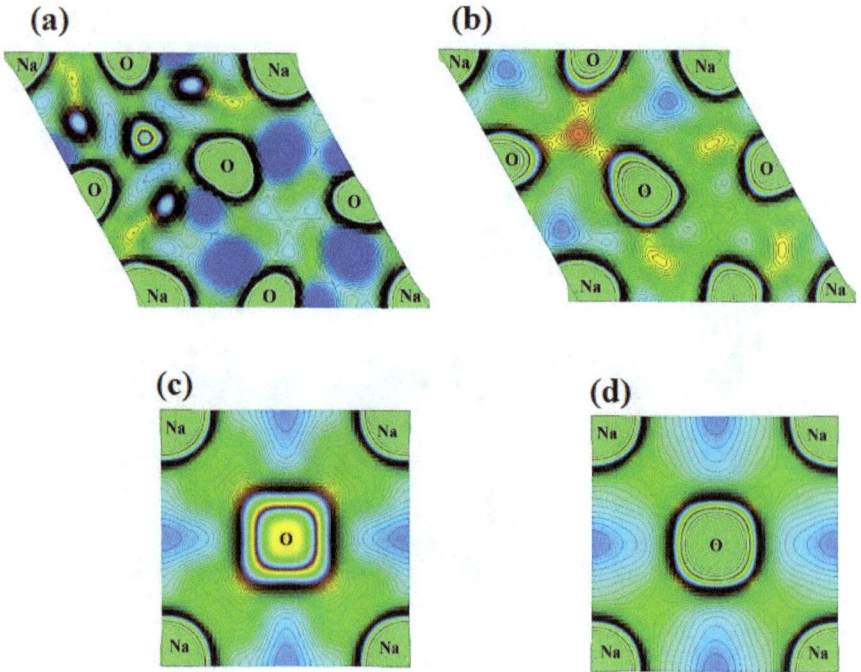

Figure 3.49 *Two dimensional charge density distributions for
$(1-x)(Na_{0.5}Bi_{0.5})TiO_3\text{-}xBaTiO_3$ ceramics on $(1\bar{2}0)$ plane for **(a)** x=0.00 **(b)** x=0.04 and
(001) plane for **(c)** x=0.08 **(d)** x=0.12.*

Lead-free Piezo-Ceramic Solid Solutions, R. Saravanan   Materials Research Forum LLC
Materials Research Foundations 41 (2018)   doi: http://dx.doi.org/10.21741/9781945291951

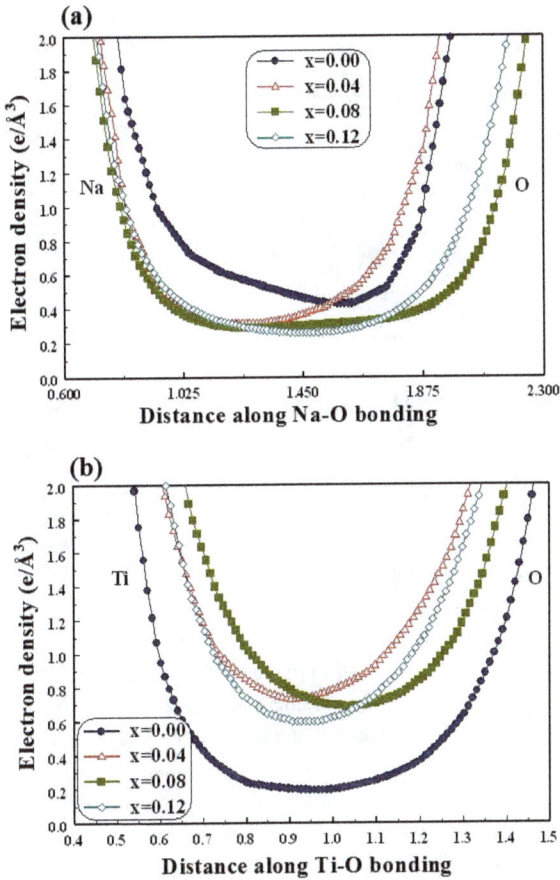

Figure 3.50 (a) One dimensional electron density profiles along [Na-O] bonding direction *(b)* One dimensional electron density profiles along [Ti-O] bonding direction.

Table 3.33 *Bond lengths and mid bond electron densities for Na-O and Ti-O bondings for $(1-x)(Na_{0.5}Bi_{0.5})TiO_3$-$xBaTiO_3$, $(x=0.00, 0.04, 0.08, 0.12)$.*

| Sample concentration (x) | Bonding | | | |
| | Na-O | | Ti-O | |
	Bond length (Å)	Mid bond electron density (e/Å³)	Bond length (Å)	Mid bond electron density (e/Å³)
x=0.00	2.5062	0.5590	1.9647	0.1919
x=0.04	2.5066	0.3139	1.9354	0.7554
x=0.08	2.7877	0.2950	1.9567	0.7072
x=0.12	2.7716	0.2650	1.9446	0.6049

3.7.4 $(1-x)(K_{0.5}Bi_{0.5})TiO_3$-$xBaTiO_3$, $(x=0.00, 0.08, 0.12)$

The three dimensional charge density distributions in the unit cell with an isosurface level of 1.0 e/Å³ for $(1-x)(K_{0.5}Bi_{0.5})TiO_3$-$xBaTiO_3$, $(x=0.00, 0.08, 0.12)$ ceramics are presented in figures 3.51 (a)-(c). Figures 3.52 (a)-(d) and 3.53 (a)-(d) show the two dimensional electron density contours in the range of 0 to 1 e/Å³ with an interval of 0.04 e/Å³. Figure 3.52 (a) shows two dimensional electron density distribution with (001) plane shaded. Figures 3.52 (b)-(d) demonstrate the two dimensional charge density contour maps corresponding to the (001) plane, which show the K/Bi-O bonds. Figure 3.53 (a) shows the three dimensional unit cell with (001) plane shaded. Figures 3.53 (b)-(d) illustrate the two dimensional contour maps corresponding to the (002) plane, which also show the Ti-O bonds. An analysis can be done on the one dimensional electron density profiles of $(1-x)(K_{0.5}Bi_{0.5})TiO_3$-$xBaTiO_3$, $(x=0.00, 0.08, 0.12)$ along K/Bi-O and Ti-O bonding directions and the valence contribution for the three systems are shown in figures 3.54 (a) and (b). The bond length variation and charge density values at mid bond of K/Bi-O and Ti-O bonds are obtained from the MEM analysis and are given in Table 3.34.

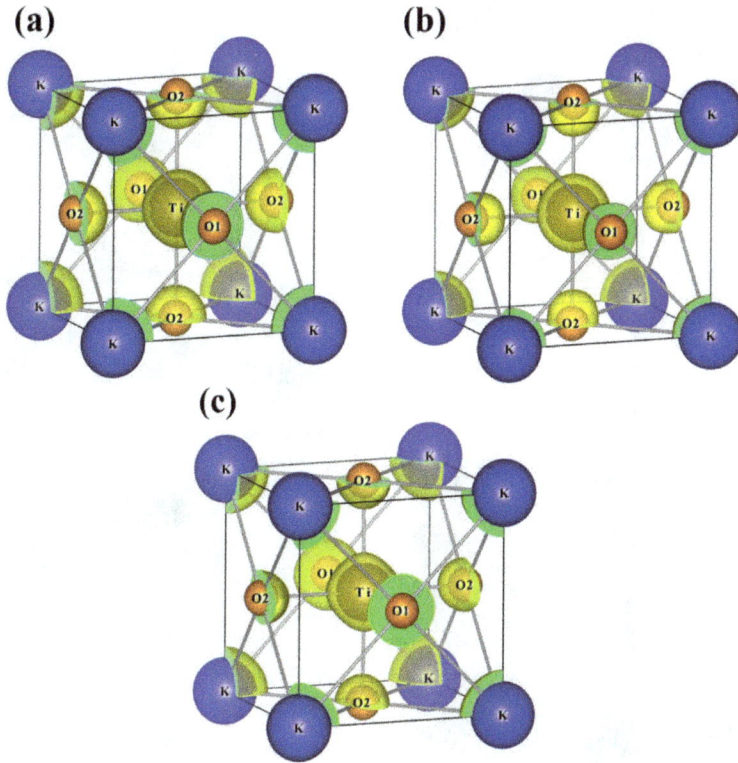

Figure 3.51 *Three dimensional charge density distribution for $(1-x)(K_{0.5}Bi_{0.5})TiO_3$-$xBaTiO_3$, **(a)** x=0.00 **(b)** x=0.08 & **(c)** x=0.12 with isosurface level of 1 e/Å³.*

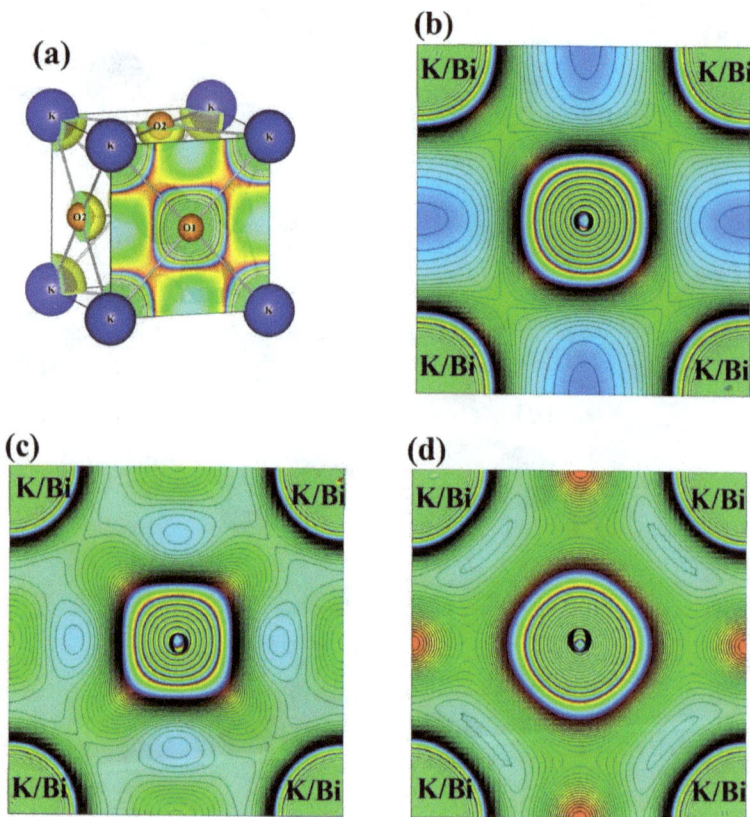

Figure 3.52 (a) *Three dimensional unit cell of* $(1-x)(K_{0.5}Bi_{0.5})TiO_3\text{-}xBaTiO_3$ *with (001) plane shaded. Two dimensional electron density distribution on (001) plane for* $(1-x)(K_{0.5}Bi_{0.5})TiO_3\text{-}xBaTiO_3,$ **(b)** *x=0.00* **(c)** *x=0.08* & **(d)** *x=0.12.*

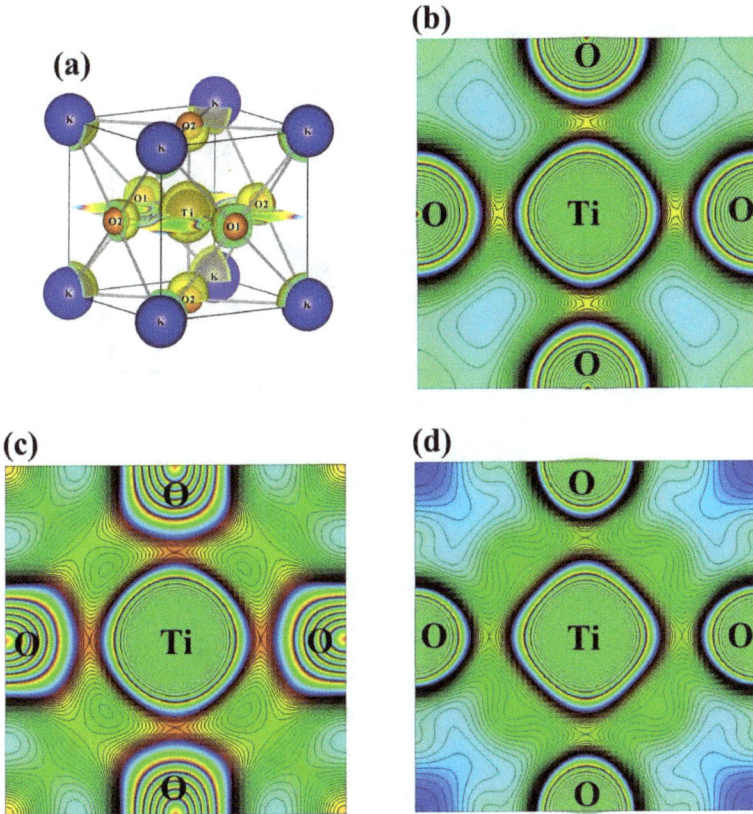

Figure 3.53 (a) Three dimensional unit cell of $(1-x)(K_{0.5}Bi_{0.5})TiO_3$-$xBaTiO_3$ with (200) plane shaded. Two dimensional electron density distribution on (200) plane for $(1-x)(K_{0.5}Bi_{0.5})TiO_3$-$xBaTiO_3$, *(b)* x=0.00 *(c)* x=0.08 & *(d)* x=0.12.

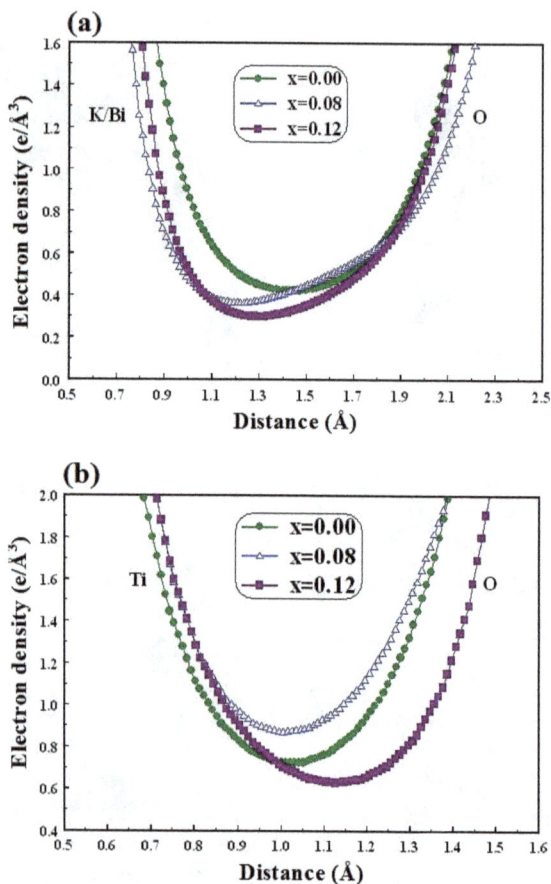

Figure 3.54 (a) One dimensional electron density profiles along K/Bi-O bonds
(b) One dimensional electron density profiles along Ti-O bonds.

Table 3.34 *Bond lengths and mid bond electron densities for K/Bi-O and Ti-O bondings for (1-x)(K$_{0.5}$Bi$_{0.5}$)TiO$_3$-xBaTiO$_3$, (x=0.00, 0.08, 0.12).*

Sample concentration (x)	Bonding			
	K/Bi-O		Ti-O	
	Bond length (Å)	Mid bond electron density (e/Å^3)	Bond length (Å)	Mid bond electron density (e/Å^3)
x=0.00	2.7880	0.4212	1.9730	0.7232
x=0.08	2.8261	0.3614	2.0011	0.8683
x=0.12	2.8265	0.2967	2.0036	0.6265

3.7.5 (1-x)Ba(Zr$_{0.2}$Ti$_{0.8}$)O$_3$-x(Ba$_{0.7}$Ca$_{0.3}$)TiO$_3$, (x=0.4, 0.5, 0.6)

Figures 3.55 (a)-(c) show the three dimensional charge density distributions in the unit cell of (1-x)Ba(Zr$_{0.2}$Ti$_{0.8}$)O$_3$-x(Ba$_{0.7}$Ca$_{0.3}$)TiO$_3$, (x=0.4, 0.5, 0.6) ceramics with similar isosurface level of 1e/Å^3. Figures 3.56 (a)-(d) display the two dimensional charge density contours in the contour range of 0-1 e/Å^3 with a contour interval of 0.04 e/Å^3 for (001) lattice plane. Figure 3.56 (a) shows the three dimensional unit cell with (001) plane shaded. Figures 3.56 (b)-(d) illustrate the two dimensional contour maps for (001) lattice plane representing the charge density distributions between Ba and O ions. Figure 3.57 (a) shows the three dimensional unit cell with (002) plane shaded. Figures 3.57 (b)-(d) illustrate the two dimensional contour maps for (002) lattice plane representing the charge density distributions between Ti and O ions. Figures 3.58 (a) and (b) represent the one dimensional electron density profiles along Ba-O and Ti-O bond respectively. The bond lengths and mid bond electron density values are given in Table 3.35.

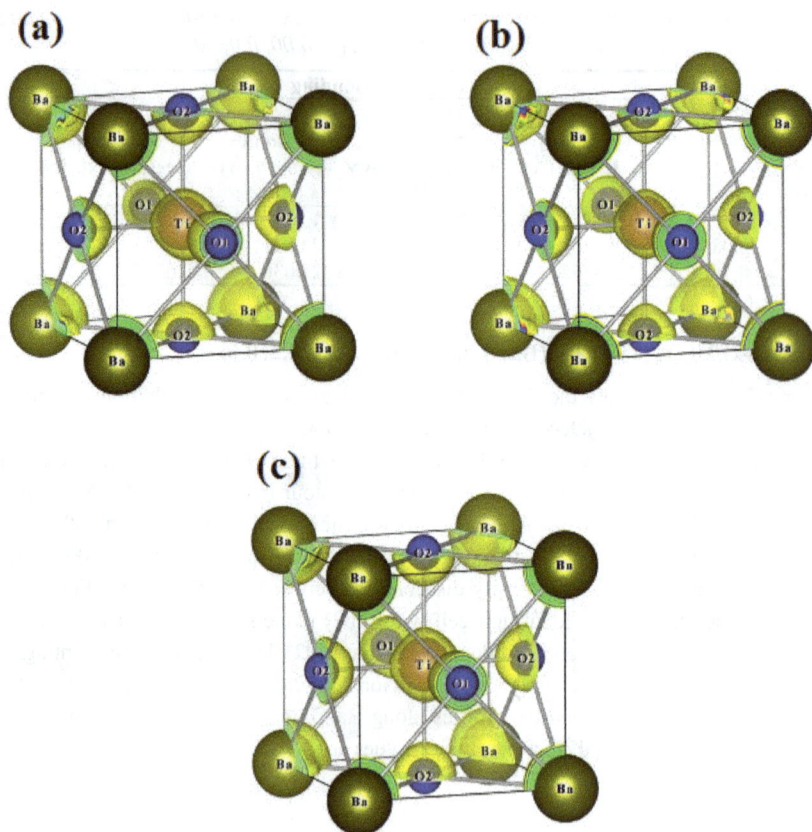

Figure 3.55 *Three dimensional charge density isosurfaces for*
(1-x)Ba(Zr$_{0.2}$Ti$_{0.8}$)O$_3$-x(Ba$_{0.7}$Ca$_{0.3}$)TiO$_3$, **(a)** *x=0.4* **(b)** *x=0.5 &* **(c)** *x=0.6.*

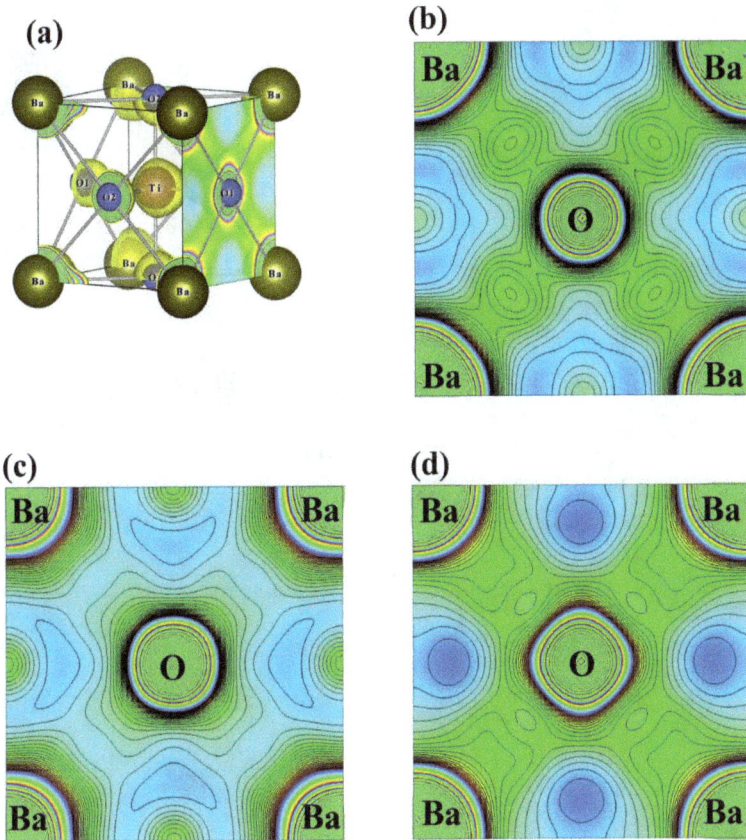

Figure 3.56 (a) Three dimensional unit cell of $(1-x)Ba(Zr_{0.2}Ti_{0.8})O_3-x(Ba_{0.7}Ca_{0.3})TiO_3$ with (001) plane shaded. Two dimensional electron density distribution on (001) plane for $(1-x)Ba(Zr_{0.2}Ti_{0.8})O_3-x(Ba_{0.7}Ca_{0.3})TiO_3$, *(b)* x=0.4 *(c)* x=0.5 & *(d)* x=0.6.

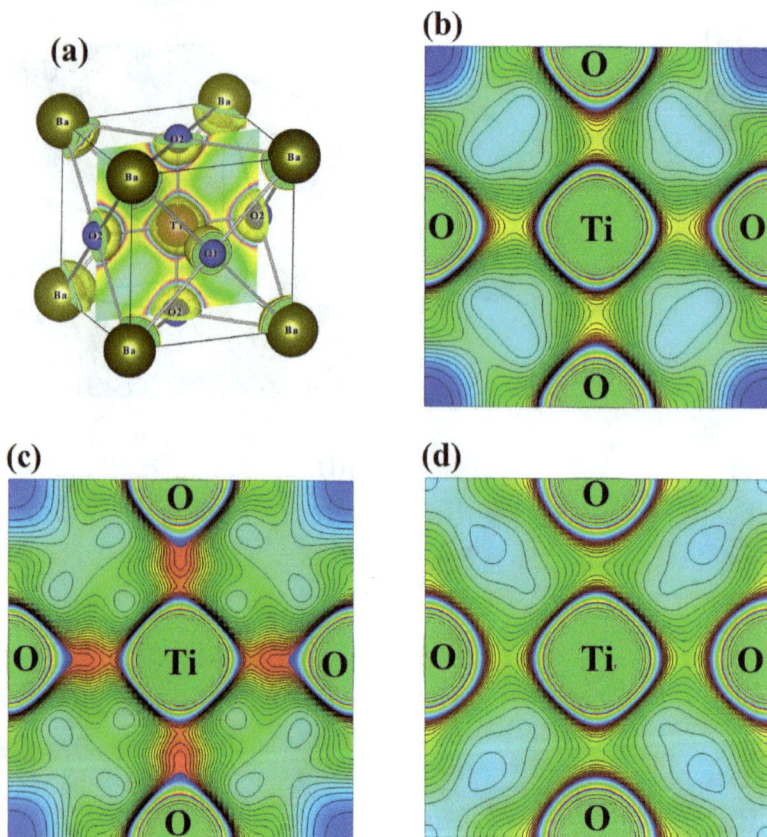

Figure 3.57 (a) *Three dimensional unit cell of $(1-x)Ba(Zr_{0.2}Ti_{0.8})O_3$-$x(Ba_{0.7}Ca_{0.3})TiO_3$ with (002) plane shaded. Two dimensional electron density distribution on (002) plane for $(1-x)Ba(Zr_{0.2}Ti_{0.8})O_3$-$x(Ba_{0.7}Ca_{0.3})TiO_3$,* ***(b)*** *x=0.4* ***(c)*** *x=0.5 &* ***(d)*** *x=0.6.*

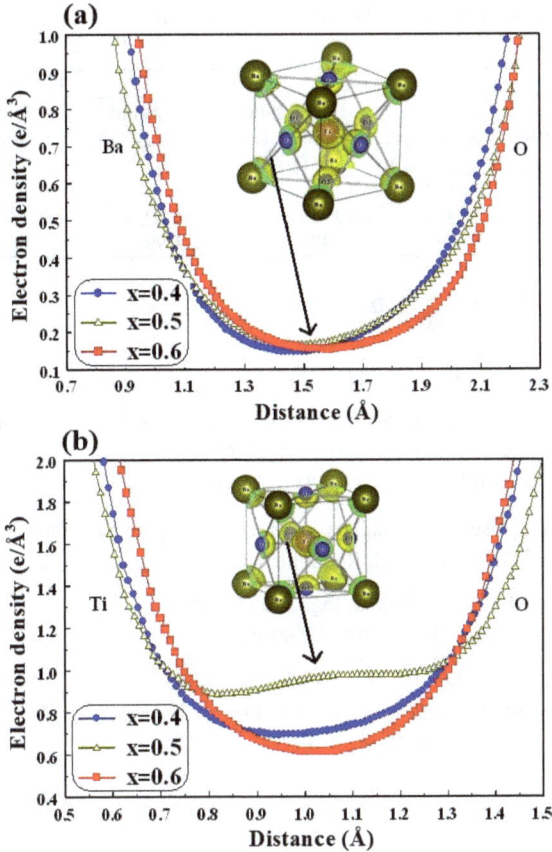

Figure 3.58 (a) *One dimensional electron density profiles along Ba-O atoms*
(b) *One dimensional electron density profiles along Ti-O bonds.*

Lead-free Piezo-Ceramic Solid Solutions, R. Saravanan   Materials Research Forum LLC
Materials Research Foundations **41** (2018)   doi: http://dx.doi.org/10.21741/9781945291951

Table 3.35 *Bond lengths and mid bond electron densities for Ba-O and Ti-O bonds for*
$(1-x)Ba(Zr_{0.2}Ti_{0.8})O_3-x(Ba_{0.7}Ca_{0.3})TiO_3$, *(x=0.4, 0.5, 0.6).*

Sample concentration (x)	Ba-O		Ti-O	
	Bond length (Å)	Mid bond electron density (e/Å³)	Bond length (Å)	Mid bond electron density (e/Å³)
x=0.4	2.8322	0.1681	2.0020	0.6998
x=0.5	2.8315	0.1689	2.0071	0.9575
x=0.6	2.8263	0.1480	1.9922	0.6150

References

[1]   Collins D.M., Nature 49, 298 (1982).

[2]   Momma K., Izumi F., VESTA 3 for three-dimensional visualization of crystal, volumetric and morphology data, J. Appl. Crystallogr. 44, 1272-1276 (2011). https://doi.org/10.1107/S0021889811038970

[3]   Petříček V., Dusek M., Palatinus J., The crystallographic computing system (Institute of Physics) (Czech Republic, Praha, 2006).

[4]   Rietveld H.M., J. Appl. Crystallogr 2, 65 (1969). https://doi.org/10.1107/S0021889869006558

[5]   Wood D.L., Tauc J., Phys. Rev. B5, 3144 (1972). https://doi.org/10.1103/PhysRevB.5.3144

Materials Research Forum LLC
doi: http://dx.doi.org/10.21741/9781945291951

Chapter 4

Analysis of results

Abstract

Chapter IV provides the detailed analysis of results obtained for all the lead-free solid solutions. Interpretation of results, comparison of physical properties and electron density distribution for all the prepared solid solutions are also discussed.

Keywords: UV-Visible Spectra, Microstructure, Correlation between Charge Density and Piezo Electric Properties, Solid Solution of $BaTiO_3$

Contents

4.1 Introduction

In this chapter, the analysis of the results obtained from various characterization methods
has been carried out. The results of structural, microstructural, optical, dielectric,
ferroelectric and piezoelectric properties of the lead-free solid solutions have been
analyzed in a detailed manner. The lead-fee solid solutions such as NKN-BT, NKNS-BT,
KBT-BT, NBT-BT, BZT-BCT powders have been prepared by solid-state reaction
method as described in chapter 2. The synthesized lead-free solid solutions have been
characterized by various techniques such as powder X-ray diffraction method (PXRD),
scanning electron microscopy (SEM), UV-visible absorption spectroscopy, energy
dispersive X-ray spectroscopy (EDS), dielectric characterizations, polarization versus
electric field hysteresis loop measurements and piezoelectric constant (d_{33})
measurements. The experimental X-ray diffraction data sets have been analyzed by
Rietveld refinement method (Rietveld, 1969) through the JANA 2006 (Petříček et al.,
2006) software. The charge density distributions in the unit cell of the prepared lead-free

solid solutions have been analyzed by maximum entropy method (MEM) (Collins, 1982), employing PRIMA (Momma and Izumi, 2011) and VESTA (Momma and Izumi, 2011) softwares.

4.2 Sample preparation

In the present book work, five series of lead-free ceramic solid solutions were prepared by solid-state reaction method. The solid solutions were prepared by mixing of two or more raw oxides and carbonates. The starting chemicals were weighed according to the stoichiometric proportions of the solid solutions to be prepared. The starting chemicals were ball milled using laboratory ball mill with agate balls. The mixed powders were calcined for several hours as given in Table 4.1. After calcination, the fine powders were pressed into circular pellets using a hydraulic press and finally the pellets were sintered using high temperature tubular furnace. The calcination temperatures, sintering temperatures and grinding durations for the synthesis of lead-free ceramics are given in Table 4.1.

Table 4.1 Calcination, sintering temperature and grinding duration for the synthesis of lead-free ceramics

Sample	Grinding duration (h)	Calcination Temp. (°C)	Calcination Duration (h)	Regrinding duration (h)	Sintering Temp. (°C)	Sintering Duration (h)
NKN-BT	2	950	4	2	1250	2
NKNS-BT	3	1000	3	3	1250	2
NBT-BT	2	850	2	2	1150	2
KBT-BT	2	800	2	2	1050	3
BCT-BZT	8	1350	2	8	1450	3

NKN-BT   - $(1-x)(Na_{1-y}K_y)NbO_3-xBaTiO_3$
NKNS-BT - $(1-x)(Na_{1-y}K_y)(Nb_{1-z}Sb_z)O_3-xBaTiO_3$
NBT-BT   - $(1-x)(Na_{0.5}Bi_{0.5})TiO_3-xBaTiO_3$
KBT-BT   - $(1-x)(K_{0.5}Bi_{0.5})TiO_3-xBaTiO_3$
BZT-BCT - $(1-x)Ba(Zr_{0.2}Ti_{0.8})O_3-x(Ba_{0.7}Ca_{0.3})TiO_3$

4.3 Powder X-ray diffraction analysis

4.3.1 $(1-x)(Na_{1-y}K_y)NbO_3-xBaTiO_3$, (x=0.1, 0.2; y=0.01, 0.05)

The X-ray diffraction patterns confirmed that the (1-x)NKN-xBT ceramics have been synthesized as single phase samples without the presence of any secondary phases (figures 3.1 (a)-(c)). It is also confirmed that $(Na_{1-y}K_y)NbO_3$ and $BaTiO_3$ ceramics form a homogenous solid solution with perovskite ABO_3 structure. The diffraction peaks shift towards lower 2θ angles with the addition of $BaTiO_3$ content. The shifting of (110) diffraction peak confirmed the incorporation of Ba^{2+} ions at (Na, K) site (figure 3.1(b)). It can be seen that the lattice constant value increases with the addition of $BaTiO_3$ (Table 3.1). This is attributed to the fact that the ionic radius of Ba^{2+} (1.34 Å) (Benlahrache *et al.*, 2004) is larger than that of Na^+ (0.97 Å) (Benlahrache *et al.*, 2004) and K^+ (1.33 Å) (Benlahrache *et al.*, 2004) ions. Due to the addition of $BaTiO_3$ (x=0.2) in the host lattice, the crystal structure of (1-x)NKN-xBT ceramics changed from tetragonal (x=0.1) to distorted cubic (x=0.2) symmetry.

At the composition x=0.1, the (200) peak splits into (200) and (002) peaks and further addition of $BaTiO_3$ (x=0.2), the diffraction peaks merge into a single (200) peak (figure 3.1 (c)). Figure 3.1 (c) confirm that the (1-x)NKN-xBT ceramics undergo a phase transition from tetragonal to distorted cubic phase. (1-x)NKN-xBT ceramics show the presence of a morphotropic phase boundary (MPB) between tetragonal and distorted cubic phase in the composition range x=0.1 to 0.2 (Zeng *et al.*, 2011). The c/a ratio of the synthesized samples decreases with addition of $BaTiO_3$ concentration as presented in Table 3.1. This also indicates that the crystal structure changes from tetragonal to distorted cubic phase. The structural refinements were carried out for (1-x)NKN-xBT ceramics using Rietveld method (Rietveld, 1969). The results obtained from the Rietveld refinement method (Rietveld, 1969) show a good agreement between the observed XRD patterns and calculated results (figures 3.2 (a)-(d)). Table 3.1 shows the variations of the lattice parameter and cell volume for different x values.

4.3.2 $(1-x)(Na_{1-y}K_y)(Nb_{1-z}Sb_z)O_3-xBaTiO_3$, (x=0.1,0.2; y=0.03,0.05; z=0.05,0.1)

The room temperature XRD patterns could be indexed according to tetragonal and cubic symmetry (figures 3.3 (a) and (b)). The shifting of the peak position towards lower angle side of 2θ is observed with the addition of $BaTiO_3$ content in the host lattice. This can be explained as follows; the ionic radius of Ba^{2+} (1.34 Å) (Shannon, 1976) is comparable to that of K^+ (1.33 Å) (Shannon, 1976) and Na^+ (0.97 Å) (Shannon, 1976) and larger than that of Nb^{5+} (0.69 Å) (Shannon, 1976). Therefore, Ba^{2+} is suitable for   A-site of the ABO_3 perovskite structure. Conversely, ionic radii of Ti^{4+} (0.68 Å) (Shannon, 1976) and Sb^{5+} (0.62 Å) (Shannon, 1976) are very close to the ionic radius of Nb^{5+} (0.69 Å) (Shannon, 1976), and hence it is introduced in B-site. Table 3.2 shows the variation of lattice constant with respect to the Ba^{2+} (x=0.1) addition, which indicates distortion in the structure of $(Na_{1-y}K_y)(Nb_{1-z}Sb_z)$ system. Furthermore, the c/a ratio was found to decrease with the increase of x and becomes unity owing to a tetragonal to cubic phase transition.

The tetragonal and cubic structures of the samples were refined in JANA 2006 (Petříček *et al.*, 2006) software by the Rietveld refinement (Rietveld, 1969) technique (figures 3.4 (a)-(d)). Various parameters including lattice parameters, scale factor, background parameter and orientation parameters were used in this technique for obtaining a minimum difference between observed and calculated profiles. Figures 3.4 (a)-(d) confirm good agreement between observed XRD patterns and theoretical fits which indicate the success of the Rietveld refinement (Rietveld, 1969) method. The good fitting is shown as the small differences near zero in the intensity scale as illustrated by a line (I_C-I_O) in the fitted XRD patterns. The cell parameters and reliability factors obtained from Rietveld refinement (Rietveld, 1969) are listed in Table 3.2. The unit cell volume of (1-x)NKNS-xBT ceramics is slightly expanded with Ba^{2+} doping as shown in Table 3.2.

4.3.3 (1-x)(Na$_{0.5}$Bi$_{0.5}$)TiO$_3$-xBaTiO$_3$, (x=0.00, 0.04, 0.08, 0.12)

The XRD results confirm pure perovskite structure and no trace of secondary phases are formed, implying that Ba^{2+} ion was successfully diffused into the $Na_{0.5}Bi_{0.5}TiO_3$ (NBT) lattice to form a solid solution (figures 3.5 (a) and (b)). All the peaks could be indexed in terms of rhombohedral structure for x=0.00 and 0.04 with space group R3c. The compositions x=0.08 and 0.12 could be indexed in terms of tetragonal structure with space group P4mm. The crystal structure of (1-x)NBT-xBT ceramics were indexed by rhombohedral symmetry at x=0.00 and 0.04, then further addition of BaTiO$_3$ (x=0.08, 0.12), the crystal structure transformed from rhombohedral to tetragonal symmetry. At x=0.08, the ceramic becomes a pure tetragonal phase. This can be evidenced by the splitting of the (202) diffraction peak into (002) and (200) peaks (figures 3.5 (b)). The splitting of the (202) peak suggests that the morphotrophic phase boundary (MPB) of rhombohedral and tetragonal phases resides at 0.04 < x < 0.08. When the quantity of BaTiO$_3$ increases in the host lattice, the positions of diffraction peaks shift towards the lower angle side of 2θ. This is attributed to the fact that the ionic radii of Na^+ (r_{Na}^+=1.39 Å) (Qu *et al.*, 2005) and Bi^{3+} (r_{Bi}^{3+}=1.28 Å) (Qu *et al.*, 2005) are smaller than that of Ba^{2+} (r_{Ba}^{2+}=1.60 Å) (Qu *et al.*, 2005), inducing lattice deformation when they enter into the A-site of host lattice.

The fitted profiles confirm that the (1-x)NBT-xBT ceramics crystallized to rhombohedral structure for x=0.00, x=0.04 with space group R3c and tetragonal for x=0.08, x=012 with space group P4mm (figures 3.6 (a)-(d)). Table 3.3 also shows the variation of lattice parameter, unit cell volume, *etc.* of (1-x)NBT-xBT ceramics. The expansions in the lattice parameter is attributed to the ionic radius of Ba^{2+} (1.60 Å) (Qu *et al.*, 2005) which is larger than those of Na^+ (1.39 Å) (Qu *et al.*, 2005) and Bi^{3+} (1.28 Å) (Qu *et al.*, 2005) ions.

4.3.4 $(1-x)(K_{0.5}Bi_{0.5})TiO_3-xBaTiO_3$, (x=0.00, 0.08, 0.12)

The experimental XRD patterns reveal the crystal structure of (1-x)KBT-xBT ceramics as tetragonal symmetry (space group: P4mm) (JCPDS PDF # 36-0339) (Chang-lin et al., 1982) (figures 3.7 (a) and (b)). The X-ray diffraction peaks shift towards lower angle side of 2θ, due to the fact that ionic radius of the substituent Ba^{2+} (1.61 Å) (Yang et al., 2016) is higher than those of K^+ (1.35 Å) (Yang et al., 2016) and Bi^{3+} (0.96 Å) (Yang et al., 2016) ions. Table 3.4 shows that the c/a ratio obtained by XRD measurement varies in the range from 1.001-1.006, which determines that the crystal structure of (1-x)KBT-xBT ceramics is tetragonal at room temperature.

The Rietveld refinements (Rietveld, 1969) were carried out by considering the positional parameters of tetragonal $BaTiO_3$ structure with space group P4mm. The initial atomic positional coordinates (x, y, z) were assumed as (0, 0, 0) for potassium/bismuth, (0.5, 0.5, 0.512) for titanium, (0.5, 0.5, 0.023) for O1 and (0.5, 0, 0.486) for O2 from standard Wyckoff (Wyckoff, 1963) crystal structure position tables. The fitting of observed and calculated profiles is quite good, confirming the tetragonal structure of (1-x)KBT-xBT ceramics (figures 3.8 (a)-(d)). The lattice constants obtained from Rietveld refinement (Rietveld, 1969) are found to be a=b=3.9760 Å, c=4.0018 Å and volume=63.23 Å (Table 3.4).

4.3.5 $(1-x)Ba(Zr_{0.2}Ti_{0.8})O_3-x(Ba_{0.7}Ca_{0.3})TiO_3$, (x=0.4, 0.5, 0.6)

From the XRD patterns, it is understood that the prepared ceramics show a single phase of perovskite structure without any trace of additional phases (figures 3.9 (a) and (b)). The crystal structures of the samples were confirmed as tetragonal (space group: P4mm) for x=0.4 and x=0.6 ceramics. The splitting of the (200) diffraction peak at 2θ~45° is observed in the composition x=0.5 as shown in figure 3.9 (b). The splitting of (200) peak suggests that the coexistence of rhombohedral and tetragonal phases (R+T) is related to the morphotropic phase boundary (MPB) region. In (1-x)BZT-xBCT system, the Ca ions prefer A-sites (Ba-sites) and Zr ions prefer B-sites (Ti-sites) in the perovskite lattice. The XRD peaks shift to higher 2θ angles with increasing x content. This is expected because the ionic radius of Ba^{2+} (1.35 Å) (Coondoo et al., 2013) is larger than that of Ca^{2+} (0.99 Å) (Coondoo et al., 2013), substituting on A-site. Similar situation exists for Zr^{4+} (0.98 Å) (Coondoo et al., 2013) substituting for Ti^{4+} (0.72 Å) (Coondoo et al., 2013) on B-site. Similar observations have also been reported by other researchers (Yingze et al., 2016; Ehmke et al., 2012).

Structural refinements were carried out for all the samples by Rietveld refinement (Rietveld, 1969) technique. The fitted refinement profiles reveal that all the samples exhibit tetragonal symmetry with space group P4mm (figures 3.10 (a)-(c)). The Ba and Ca atoms are assumed to have shared the same crystallographic A-site of the perovskite structure. Similarly, Ti and Zr atoms are positioned at B-site. Results of the structural refinement show that the diffraction peaks of x=0.4 and 0.6 can be indexed to tetragonal phase with space group P4mm. Figure 3.10 (b) shows the mixed phase refinement profile for the MPB composition 0.5BZT-0.5BCT with a predominant tetragonal phase (P4mm) and a weak rhombohedral phase (R3m). The structural parameter, reliability indices and goodness of fit are summarised in Table 3.5.

Table 4.2 Comparison of lattice constants for the synthesized lead-free solid solutions

Sample	Concentration	Phase	a (Å)	b (Å)	c (Å)
NKN-BT	x=0.1, y=0.01	Tetragonal	3.9250 (4)	3.9250 (4)	3.9587(13)
	x=0.1, y=0.05	Tetragonal	3.9271(5)	3.9271(5)	3.9642(4)
	x=0.2, y=0.01	Cubic	3.9562(6)	3.9562(6)	3.9672(6)
	x=0.2, y=0.05	Cubic	3.9613(11)	3.9613(11)	3.9763(11)
NKNS-BT	x=0.1, y=0.03, z=0.05	Tetragonal	3.9286(6)	3.9286(6)	3.9387(6)
	x=0.1, y=0.05, z=0.05	Tetragonal	3.9339(8)	3.9339(8)	3.9420(8)
	x=0.2, y=0.03, z=0.1	Cubic	3.9389(3)	3.9389(3)	3.9389(3)
	x=0.2, y=0.05, z=0.1	Cubic	3.9442(5)	3.9442(5)	3.9442(5)
NBT-BT	x=0.00	Rhombohedral	5.4746(9)	5.4746(9)	13.5207(8)
	x=0.04	Rhombohedral	5.4809(16)	5.4809(16)	13.5322(14)
	x=0.08	Tetragonal	3.8746(12)	3.8746(12)	3.9226(9)
	x=0.12	Tetragonal	3.8879(2)	3.8879(2)	3.9582(2)
KBT-BT	x=0.00	Tetragonal	3.9760(6)	3.9760(6)	4.0018(6)
	x=0.08	Tetragonal	4.0139(3)	4.0139(3)	4.0201(3)
	x=0.12	Tetragonal	4.0287(2)	4.0287(2)	4.0541(2)
BZT-BCT	x=0.4	Tetragonal	4.0040(1)	4.0040(1)	4.0065 (2)
	x=0.5	Rhombohedral + Tetragonal	4.0141(1) 5.4455(2)	4.0141(1) 5.4455(2)	3.9945(1) 6.6334(3)
	x=0.6	Tetragonal	3.9843(5)	3.9843(5)	4.0097(4)

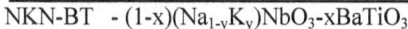

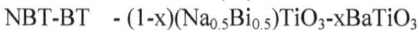

NKN-BT - $(1-x)(Na_{1-y}K_y)NbO_3-xBaTiO_3$
NKNS-BT - $(1-x)(Na_{1-y}K_y)(Nb_{1-z}Sb_z)O_3-xBaTiO_3$
NBT-BT - $(1-x)(Na_{0.5}Bi_{0.5})TiO_3-xBaTiO_3$
KBT-BT - $(1-x)(K_{0.5}Bi_{0.5})TiO_3-xBaTiO_3$
BZT-BCT - $(1-x)Ba(Zr_{0.2}Ti_{0.8})O_3-x(Ba_{0.7}Ca_{0.3})TiO_3$

The X-ray diffraction pattern and Rietveld refinement (Rietveld, 1969) analysis revealed that all the prepared ceramics crystallize without any secondary phase and the calculated peak intensities match well with the observed ones. The refined lattice constant values for all the lead-free solid solutions are compared in Table 4.2. Table 4.2 shows the variation of the lattice constant values corresponding to addition of $BaTiO_3$ dopant in the host lattice. The cell constant values increase with the influence of the $BaTiO_3$ addition. This is due to the ionic radius of Ba^{2+} is higher than the ionic radius of ions substituted through solid-state reaction (SSR).

4.4   Microstructure and elemental analysis of all the lead-free solid solutions

The scanning electron microscopic images were recorded by surface scanning of the pellets. The SEM images of $(1-x)(Na_{1-y}K_y)NbO_3-xBaTiO_3$, (x=0.1, 0.2; y=0.01, 0.05) ceramics show that the particles are uniformly distributed without much agglomeration (figures 3.11 (a)-(d)). The average particle size gradually decreases with the addition of $BaTiO_3$ content (Table 3.6). The EDS results of $(1-x)(Na_{1-y}K_y)NbO_3-xBaTiO_3$ ceramics indicate that the constituent ions are present in the respective samples with expected proportions (figures 3.12 (a)-(d)). No additional elements are detected through EDS spectrum. The numerical values of atomic and weight percentages show that the preparation condition completely favours the formation of $(1-x)(Na_{1-y}K_y)NbO_3-xBaTiO_3$ ceramics (Table 3.7 (a) and (b)).

The SEM images of lead-free $(1-x)(Na_{1-y}K_y)(Nb_{1-z}Sb_z)O_3-xBaTiO_3$, (x=0.1, 0.2; y=0.03, 0.05; z=0.05, 0.1) ceramics show for the composition x=0.1, y=0.03, z=0.05 (figure 3.13 (a)), particles with porous structure and for the composition x=0.1, y=0.05, z=0.05 (figures 3.13 (b)), the particles are almost spherical in shape with less voids on the surface. At the composition x=0.2, the morphology of the sample shows aggregated square shape particles with low porosities. The variation in average particle size of $(1-x)(Na_{1-y}K_y)(Nb_{1-z}Sb_z)O_3-xBaTiO_3$ ceramics is observed with the addition of $BaTiO_3$ content (figures 3.13 (a)-(d)). The average particle sizes of the prepared samples decrease with the addition of $BaTiO_3$ content (Table 3.8). The EDS spectra of $(1-x)(Na_{1-y}K_y)(Nb_{1-z}Sb_z)O_3-xBaTiO_3$ ceramics show the presence of Na, Nb, Sb, K, Ba, Ti, and O ions in the ceramics (figures 3.14 (a)-(d)). The atomic and weight percentages of elements indicate that the Ba^{2+} has been incorporated into $(Na_{1-y}K_y)(Nb_{1-z}Sb_z)O_3$ lattice (Tables 3.9 (a) and (b)).

From the SEM micrographs, it is found that the size and shape of the powder particles of (1-x)NBT-xBT ceramics apparently change with the addition of $BaTiO_3$ content (figures 3.15 (a)-(d)). This suggests that $BaTiO_3$ has dissolved in $(Na_{0.5}Bi_{0.5})TiO_3$ and it in turn inhibits the growth of the particles. The average particle sizes of the (1-x)NBT-xBT ceramics are tabulated in Table 3.10. The EDS spectra of (1-x)NBT-xBT ceramics confirm that the preparation condition favours the formation of the solid solutions (figures 3.16 (a)-(d)). The EDS results indicate that the constituent ions are present in the

Materials Research Forum LLC
doi: http://dx.doi.org/10.21741/9781945291951

respective samples in expected proportions. It is observed that the stoichiometries of the elements present in the solid solution match well with their nominal stoichiometries (Table 3.11).

It can be seen that there is a significant change in the morphology and particle size of the $(1-x)$KBT-xBT ceramics with the addition of $BaTiO_3$ content (figures 3.17 (a)- (c)). Particles of irregular shape with pores heterogeneously distributed are seen at x=0.08 composition and very fine particles with agglomeration is noticed at x=0.12 (figures 3.17 (b) and (c)). The average particle sizes of the prepared samples are given in Table 3.12. Each peak in the EDS spectra refers to particular atomic species present in the prepared sample (figures 3.18 (a)-(c)). The atomic and weight percentages of the elements (K, Bi, Ti, Ba and O) present in the synthesized sample with the expected proportions are given in Table 3.13. It confirmed that there are no impurities present in the sample.

Table 4.3 *Comparison of average particle sizes for synthesized lead-free solid solutions.*

Sample	Concentration	Average particle size (μm)
NKN-BT	x=0.1, y=0.01	0.81
	x=0.1, y=0.05	0.65
	x=0.2, y=0.01	0.78
	x=0.2, y=0.05	0.96
NKNS-BT	x=0.1, y=0.03, z=0.05	1.5
	x=0.1, y=0.05, z=0.05	0.53
	x=0.2, y=0.03, z=0.1	0.46
	x=0.2, y=0.05, z=0.1	0.96
NBT-BT	x=0.00	1.53
	x=0.04	1.85
	x=0.08	3.20
	x=0.12	1.30
KBT-BT	x=0.00	0.81
	x=0.08	0.61
	x=0.12	0.43
BZT-BCT	x=0.4	1.14
	x=0.5	2.05
	x=0.6	1.57

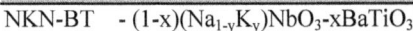

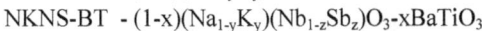

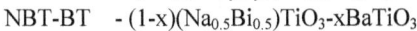

NKN-BT - $(1-x)(Na_{1-y}K_y)NbO_3-xBaTiO_3$
NKNS-BT - $(1-x)(Na_{1-y}K_y)(Nb_{1-z}Sb_z)O_3-xBaTiO_3$
NBT-BT - $(1-x)(Na_{0.5}Bi_{0.5})TiO_3-xBaTiO_3$
KBT-BT - $(1-x)(K_{0.5}Bi_{0.5})TiO_3-xBaTiO_3$
BZT-BCT - $(1-x)Ba(Zr_{0.2}Ti_{0.8})O_3-x(Ba_{0.7}Ca_{0.3})TiO_3$

In (1-x)BZT-xBCT ceramics, the microstructure of the composition x=0.4 exhibited rectangular-like grains as shown in figure 3.19 (b). The shape of the particles changes with respect to the Ca^{2+} concentration. The average particle sizes of the prepared samples are found to increase with the addition of in Ca^{2+} concentration (Table 3.14). The presence of all elements in these ceramics detected by EDS spectra are listed in Table 3.15. The EDS results reveal that the observed atomic and weight percentages match well with expected values.

The average particle sizes for all the lead-free ceramics are compared in Table 4.3.

4.5 UV-visible spectra analysis of all the lead-free solid solutions

The optical band gap energy values of the (1-x)NKN-xBT (Table 3.16) and (1-x)NKNS-xBT (Table 3.17) ceramics decrease with the addition of BaTiO$_3$ content. It is concluded that during the substitution of BaTiO$_3$, the carrier concentration increases and consequently the band gap is reduced. The band gap values for the prepared (1-x)NBT-xBT system range from 3.027 eV to 3.142 eV (Table 3.18). Table 3.18 shows that the optical band gap of the (1-x)NBT-xBT ceramics increases with the addition of BaTiO$_3$ concentration. The optical band gap values are closer to the reported values (Sridevi *et al.*, 2015).

The direct band gap of (1-x)KBT-xBT, (x=0.00, 0.08, 0.12) ceramics varies from 3.006 eV to 3.088 eV (Table 3.19). The band gap energy (E_g) of (1-x)KBT-xBT ceramics was determined by extrapolating the straight line portion of the curve to the photon energy axis (Table 3.19). The optical band gap values for the (1-x)BZT-xBCT, (x=0.4, 0.5, 0.6) ceramics are given in Table 3.20. The band gap values obtained for all the samples range from 3.15 eV to 3.07 eV, which are slightly smaller than reported (Rani *et al.*, 2014) value, 3.26 eV.

The optical band gap values estimated from Tauc's relation for all the lead-free solid solutions and tabulated in Table 4.4.

Table 4.4 *Comparison of band gap energy values for synthesized lead-free solid solutions.*

Sample	Concentration	Band gap (eV)
NKN-BT	x=0.1, y=0.01	3.447
	x=0.1, y=0.05	3.373
	x=0.2, y=0.01	3.421
	x=0.2, y=0.05	3.392
NKNS-BT	x=0.1, y=0.03, z=0.05	3.343
	x=0.1, y=0.05, z=0.05	3.177
	x=0.2, y=0.03, z=0.1	3.255
	x=0.2, y=0.05, z=0.1	3.243
NBT-BT	x=0.00	3.027
	x=0.04	3.084
	x=0.08	3.110
	x=0.12	3.142
KBT-BT	x=0.00	3.006
	x=0.08	3.050
	x=0.12	3.088
BZT-BCT	x=0.4	3.154
	x=0.5	3.137
	x=0.6	3.074

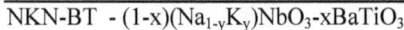

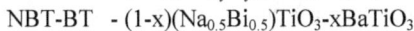

NKN-BT   - $(1\text{-}x)(Na_{1\text{-}y}K_y)NbO_3\text{-}xBaTiO_3$
NKNS-BT- $(1\text{-}x)(Na_{1\text{-}y}K_y)(Nb_{1\text{-}z}Sb_z)O_3\text{-}xBaTiO_3$
NBT-BT   - $(1\text{-}x)(Na_{0.5}Bi_{0.5})TiO_3\text{-}xBaTiO_3$
KBT-BT   - $(1\text{-}x)(K_{0.5}Bi_{0.5})TiO_3\text{-}xBaTiO_3$
BZT-BCT - $(1\text{-}x)Ba(Zr_{0.2}Ti_{0.8})O_3\text{-}x(Ba_{0.7}Ca_{0.3})TiO_3$

4.6   Dielectric properties of all the lead-free solid solutions

The temperature dependent dielectric constant (ε) and dielectric loss (tan δ) of (1-x)NKN-xBT ceramics were measured at 1kHz (figures 3.26 and 3.27). A sharp dielectric peak is observed at 280°C for the composition x=0.1, which denotes a typical ferroelectric to paraelectric phase transition. On further addition of BaTiO$_3$ (x=0.2), the phase transition temperature increases. Table 3.21 shows the dielectric constant (ε) and dielectric loss (tan δ) of (1-x)NKN-xBT ceramics at 1 kHz.

In the case of (1-x)NKNS-xBT ceramics, a phase transition peak is observed at 120°C, corresponding to the tetragonal-cubic transition (figures 3.28 (a)-(d)). On further increasing the x value, dielectric constant (ε) is suppressed and the peak disappears (Zeng et al., 2011). The phase transition from tetragonal to cubic structure leads to a reduction

Lead-free Piezo-Ceramic Solid Solutions, R. Saravanan   Materials Research Forum LLC
Materials Research Foundations **41** (2018)   doi: http://dx.doi.org/10.21741/9781945291951

in the ferroelectric transition temperature and the value of the dielectric constant (ε). The temperature dependent dielectric loss (tan δ) for (1-x)NKNS-xBT ceramics were measured at different frequencies (figures 3.29 (a)-(d)).

In (1-x)NBT-xBT ceramics, the maximum dielectric constant (ε) is observed at x=0.08 (Table 3.23). The corresponding temperature for maximum dielectric constant (ε) of undoped $Na_{0.5}Bi_{0.5}TiO_3$ is observed at around 320°C. With the addition of $BaTiO_3$ in the host lattice, the dielectric plot for x=0.08 shows intense phase transition peak at 280°C (figures 3.30 (a)-(d)). This can be attributed to the fact that the structural transition takes place from rhombohedral to tetragonal phase. Thus, the MPB composition enables the dipole moments to align efficiently with the field, resulting in a higher polarizability of the material (Bhattacharya and Ravichandran, 2003).

The dielectric constant (ε) of (1-x)KBT-xBT ceramics decreases with increasing frequency and becomes constant above 1 kHz (Figure 3.32 (a)). This is attributed to the different types of polarizations (electronic, atomic, interfacial and ionic, *etc.*) may exist at lower frequencies of the dielectric plot. At higher frequencies, the dielectric constant (ε) decreases due to the contribution from electronic polarization (Koops, 1951; Patil *et al.*, 2007). The variation of dielectric loss (tan δ) as a function of frequency at room temperature for (1-x)KBT-xBT ceramics is shown in figure 3.32 (b). A decreasing trend is observed in loss values at higher frequencies with the addition of $BaTiO_3$ in the host lattice (Table 3.24). The dielectric constant (ε) and loss (tan δ) for undoped $K_{0.5}Bi_{0.5}TiO_3$ ceramics are found to be 511 and 0.51 respectively at 10 kHz frequency measured at room temperature (Table 3.24).

In (1-x)BZT-xBCT ceramics, the Curie temperature (T_C) shift towards higher temperature side with the increase of x value. The highest dielectric constant (ε~3702) is observed for the composition 0.5BZT-0.5BCT, due to the presence of mixed structures (Table 3.25). The coexistence of rhombohedral and tetragonal phases increases the number of orientations (in polarization) in this composition, which leads to higher dielectric constant.

A comparison of dielectric constant values for all prepared lead-free ceramics is given in Table 4.5. The maximum values of dielectric constant (ε) are observed near the morphotropic phase boundary. The dissipation factor is defined as the tangent of the loss angle (tan δ). It is a measure of the amount of electrical energy which is lost through conduction when a voltage is applied across the ceramic sample. Table 4.5 shows the dielectric loss values of the all prepared lead-free ceramics.

Table 4.5 *Comparison of dielectric constant (ε) and dielectric loss (tan δ) values of synthesized lead-free solid solutions.*

Sample	Concentration	Phases	Dielectric constant (ε)	Dielectric loss (tan δ)
NKN-BT	x=0.1, y=0.01	Tetragonal	1528	0.375
	x=0.1, y=0.05	Tetragonal	5002	1.855
	x=0.2, y=0.01	Cubic	3547	1.218
	x=0.2, y=0.05	Cubic	3031	0.960
NKNS-BT	x=0.1, y=0.03, z=0.05	Tetragonal	4744	0.026
	x=0.1, y=0.05, z=0.05	Tetragonal	3161	0.028
	x=0.2, y=0.03, z=0.1	Cubic	650	0.004
	x=0.2, y=0.05, z=0.1	Cubic	435	0.011
NBT-BT	x=0.00	Rhombohedral	2199	0.163
	x=0.04	Rhombohedral	3500	0.111
	x=0.08	Tetragonal	4070	0.082
	x=0.12	Tetragonal	779	0.057
KBT-BT	x=0.00	Tetragonal	511	0.512
	x=0.08	Tetragonal	418	0.260
	x=0.12	Tetragonal	529	0.301
BZT-BCT	x=0.4	Tetragonal	3374	-
	x=0.5	Rhombohedral + Tetragonal	3702	-
	x=0.6	Tetragonal	2151	-

NKN-BT   - $(1-x)(Na_{1-y}K_y)NbO_3-xBaTiO_3$

NKNS-BT - $(1-x)(Na_{1-y}K_y)(Nb_{1-z}Sb_z)O_3-xBaTiO_3$

NBT-BT   - $(1-x)(Na_{0.5}Bi_{0.5})TiO_3-xBaTiO_3$

KBT-BT   - $(1-x)(K_{0.5}Bi_{0.5})TiO_3-xBaTiO_3$

BZT-BCT - $(1-x)Ba(Zr_{0.2}Ti_{0.8})O_3-x(Ba_{0.7}Ca_{0.3})TiO_3$

4.7   Ferroelectric and piezoelectric properties

4.7.1 $(1-x)(Na_{1-y}K_y)NbO_3-xBaTiO_3$, (x=0.1, 0.2; y=0.01, 0.05)

The ferroelectric properties of (1-x)NKN-xBT ceramics were analyzed by polarization versus electric field hysteresis loops (figures 3.34 (a)-(d)). The saturated hysteresis loop is observed at x=0.1 composition, which also crystallized with tetragonal structure. With the addition of $BaTiO_3$ (x=0.2) content, the hysteresis loops with low remnant polarization (P_r) and high coercive filed (E_C) are observed (figures 3.34 (a)-(d)). The reduction in the ferroelectric property is attributed to the phase transition from tetragonal to distorted cubic. The remnant polarizations for the prepared samples (x=0.1) closely agree with the previously reported values (Zeng *et al.*, 2011) (Table 3.26). From Table

3.26, it is understood that the incorporation of a small amount (x=0.1) of BaTiO$_3$ improves the piezoelectric properties of (1-x)NKN-xBT ceramics. The maximum piezoelectric constant (d$_{33}$) is obtained for x=0.1, which also shows largest remnant polarization and lowest coercive field. The largest remnant polarization and lowest coercive field imply an essential relation between the piezoelectric property and the ferroelectric nature of the (1-x)NKN-xBT ceramics. At x=0.2, the crystal structure of (1-x)NKN-xBT changes to distorted cubic and shows lower remnant polarization. The lower value of remnant polarization is attributed to the reduction of piezoelectric property at x=0.2 composition.

4.7.2 (1-x)(Na$_{1-y}$K$_y$)(Nb$_{1-z}$Sb$_z$)O$_3$-xBaTiO$_3$, (x=0.1,0.2; y=0.03,0.05; z=0.05,0.1)

In (1-x)NKNS-xBT ceramics, the samples with composition up to x=0.1 show a similar hysteresis loop confirming ferroelectric behaviour (figures 3.35 (a)-(b)). This is attributed to the internal distortion of the BO$_6$ octahedron in the unit cell of (1-x)NKNS-xBT ceramics (Shah and Kotnala, 2013). It can be seen that the value of maximum polarization (P$_m$) and remnant polarization (P$_r$) is enhanced at x=0.1 (Table 3.27). The ferroelectric properties of (1-x)NKNS-xBT ceramics decrease with increasing BaTiO$_3$ (x=0.2) content due to high coercive field. There is a significant decrease in the values of remnant polarization in higher doping of BaTiO$_3$ (x=0.2) (Table 3.27). It can be seen that the addition of BaTiO$_3$ content in the host lattice reduces the remnant polarization (P$_r$) drastically from 14.86 μC/cm^2 to 2.86 μC/cm^2 (Table 3.27). This is attributed to the transformation of the tetragonal crystal structure to cubic structure.

The (1-x)NKNS-xBT ceramics were polarized by choosing the poling temperature below the phase transition temperature of the ceramics. Therefore, the optimized poling conditions were used to investigate the piezoelectric activity of (1-x)NKNS-xBT ceramics. In general, the low poling electric field makes the polarization switching inadequate. However, the excessive poling electric field tends to over-pole the samples which lead to physical flaws and eventually to the dielectric breakdown of the samples. The optimum poling electric field for (1-x)NKNS-xBT ceramics is about two times that of E$_C$ (Fu *et al.*, 2010). The poling temperature and time were fixed as constants and the piezoelectric coefficient (d$_{33}$) were measured as a function of applied d.c field (Table 3.27). It is evident that the incorporation of a small amount (x=0.1) of BaTiO$_3$ improves the piezoelectric properties of (1-x)NKNS-xBT ceramics and the higher substitution of BaTiO$_3$ suggests less piezoelectric property. This is attributed to the phase transition from tetragonal to cubic structure.

4.7.3 $(1-x)(Na_{0.5}Bi_{0.5})TiO_3-xBaTiO_3$, (x=0.00, 0.04, 0.08, 0.12)

In the case of (1-x)NBT-xBT ceramics, a saturated hysteresis loop was observed at x=0.00 (figure 3.36 (a)). Higher remnant polarization ($P_r \sim 18.92$ $\mu C/cm^2$) with reduced coercive field ($E_C \sim 22.68$ kV/cm) is obtained in x=0.08 ceramics (Table 3.28). This confirms the near MPB nature of this composition. The low value of coercive field (E_C) in x=0.08 composition is attributed to the enhancement of the ferroelectric and piezoelectric properties. Since, with lower E_C value, the domain movement can take place easily in the unit cell and can result in enhanced piezoelectric properties. With the further addition of $BaTiO_3$ in the host lattice, a low remnant polarization (P_r) value is observed at x=0.12 (Table 3.28). This is attributed to the lower lattice distortion of the tetragonal symmetry. The highest value of piezoelectric coefficient is obtained for x=0.08 composition which also confirms the near MPB nature. The sample x=0.08 exhibits higher d_{33} value, due to the presence of large number of poling directions (Table 3.28). Similar results have also been observed by Xu et al., (2008) (Xu et al., 2008).

4.7.4 $(1-x)(K_{0.5}Bi_{0.5})TiO_3-xBaTiO_3$, (x=0.00, 0.08, 0.12)

A nearly saturated hysteresis loop was found for undoped $K_{0.5}Bi_{0.5}TiO_3$ ceramic under an electric field of E=6 kV/cm (figure 3.37 (a)). Enhanced ferroelectric property is observed in undoped $K_{0.5}Bi_{0.5}TiO_3$ ceramic due to the lone pair effect of electron in the Bi^{3+}. The values of coercive field (E_C), remnant polarization (P_r) and maximum polarization (P_m) measured at room temperature are given in Table 3.29.

4.7.5 $(1-x)Ba(Zr_{0.2}Ti_{0.8})O_3-x(Ba_{0.7}Ca_{0.3})TiO_3$, (x=0.4, 0.5, 0.6)

In (1-x)BZT-xBCT ceramics, enhanced ferroelectric properties are obtained at the 0.5BZT-0.5BCT composition measured at room temperature, implying an MPB like behaviour as suggested by XRD patterns (figure 3.38 (b)). The maximum piezoelectric constant (d_{33}) is observed in 0.5BZT-0.5BCT ceramics. The d_{33} value is found to be maximum at the composition x=0.5 (d_{33}=276 pC/N), which is attributed to the coexistence of rhombohedral and tetragonal phases (Table 3.30). The presence of mixed phase in the composition x=0.5 provides favorable condition for easier rotation of the polarization vectors (Mishra et al., 2014; Shi et al., 2011). The existence of MPB with rhombohedral and tetragonal phases in 0.5BZT-0.5BCT ceramics plays a most important role in improving the piezoelectric properties. Similar results have also been reported in literature (Mishra et al., 2014).

The samples exhibit ferroelectric nature within the applied field without dielectric breakdown. The ferroelectric and piezoelectric properties for all prepared lead-free ceramics are compared in Table 4.6. Among the prepared lead-free solid solution ceramics, the series (1-x)BZT-xBCT has the highest values of piezoelectric constant (d_{33}). In particular in this series, the sample 0.5BZT-0.5BCT has the higher d_{33} (276 pC/N) value.

Table 4.6 *Comparison of ferroelectric and piezoelectric properties of synthesized lead-free solid solutions.*

Sample	Concentration	P_m ($\mu C/cm^2$)	P_r ($\mu C/cm^2$)	E_C (kV/cm)	d_{33} (pC/N)
NKN-BT	x=0.1, y=0.01	16.32	12.88	15.73	120
	x=0.1, y=0.05	14.91	11.56	14.86	104
	x=0.2, y=0.01	3.82	2.86	15.78	38
	x=0.2, y=0.05	4.38	3.22	16.91	34
NKNS-BT	x=0.1, y=0.03, z=0.05	16.73	14.83	11.47	110
	x=0.1, y=0.05, z=0.05	12.49	9.28	8.29	94
	x=0.2, y=0.03, z=0.1	3.81	2.86	15.72	54
	x=0.2, y=0.05, z=0.1	4.43	3.74	3.35	35
NBT-BT	x=0.00	21.45	20.60	28.77	63
	x=0.04	14.84	14.09	33.81	78
	x=0.08	20.24	18.92	22.68	122
	x=0.12	2.43	0.53	5.67	58
KBT-BT	x=0.00	51.51	51.95	5.22	-
	x=0.08	5.58	5.24	16.64	-
	x=0.12	5.01	3.47	15.03	-
BZT-BCT	x=0.4	14.37	5.98	6.02	215
	x=0.5	15.12	4.90	3.62	276
	x=0.6	14.06	5.76	4.88	176

P_m - maximum polarization
P_r - remnant polarization
E_C - coercive field
d_{33} - piezoelectric coefficient
NKN-BT   - $(1-x)(Na_{1-y}K_y)NbO_3-xBaTiO_3$
NKNS-BT - $(1-x)(Na_{1-y}K_y)(Nb_{1-z}Sb_z)O_3-xBaTiO_3$
NBT-BT   - $(1-x)(Na_{0.5}Bi_{0.5})TiO_3-xBaTiO_3$
KBT-BT   - $(1-x)(K_{0.5}Bi_{0.5})TiO_3-xBaTiO_3$
BZT-BCT - $(1-x)Ba(Zr_{0.2}Ti_{0.8})O_3-x(Ba_{0.7}Ca_{0.3})TiO_3$

Lead-free Piezo-Ceramic Solid Solutions, R. Saravanan
Materials Research Foundations **41** (2018)

Materials Research Forum LLC
doi: http://dx.doi.org/10.21741/9781945291951

4.8 Charge density analysis

4.8.1 (1-x)(Na$_{1-y}$K$_y$)NbO$_3$-xBaTiO$_3$, (x=0.1, 0.2; y=0.01, 0.05)

It is seen from figures 3.39 (a)-(d), the three dimensional structures show the distribution of charges between constituent atoms inside the unit cell. The two dimensional electron density distributions on (001) plane shows the bonding between Na and O atoms (figures 3.40 (a)-(d)). The contours lines between Na and O atoms fade away from the boundary of the two atoms which reveals that there is no charge accumulation at the middle of the bond (figure 3.40 (a)-(d)). This confirms that Na-O bond exhibits ionic nature. But, as far as the two dimensional electron density map for the Nb-O bond in the (002) plane is concerned, it is clear from the figures 3.41 (a)-(d) that there is an increase of charge in the bonding region between the two atoms which supports that Nb-O bond is covalent in nature. These results are quantitatively analyzed by plotting one dimensional charge density profiles along Na-O and Nb-O bond (figures 3.42 (a) and (b)). The bond length values and mid bond electron density values are given in Table 3.31.

4.8.1.1 Correlation between charge density and piezoelectric properties of (1-x)NKN-xBT ceramics

The piezoelectric response of the lead-free (1-x)(Na$_{1-y}$K$_y$)NbO$_3$-xBaTiO$_3$, (x=0.1, 0.2; y=0.01, 0.05) ceramics were correlated with charge density properties. For the compositions x=0.1; y=0.01 and x=0.1; y=0.05, the mid bond electron density values for Nb-O bond are 1.0558 e/Å^3 and 0.6317 e/Å^3, respectively (Table 3.31). These higher values of mid bond electron density confirmed the covalent character between Nb and O atoms. The presence of more charges in the mid bond region of Nb-O atoms is attributed to the enhanced ferroelectric and piezoelectric properties. With the addition of BaTiO$_3$ (x=0.2) on (Na$_{1-y}$K$_y$)NbO$_3$ ceramics, the maximum polarization (P$_m$) value decreases to 3.82 μC/cm^2. This is confirmed from the XRD results also that the tetragonal crystal structure transforms to distorted cubic structure. The phase change of the crystal structure is attributed to the reduction in displacement of off centered octahedral Nb ion. This may lead to reduction of ferroelectric and piezoelectric properties in x=0.2 composition. The distorted cubic structure (x=0.2) also has spontaneous polarization (due to distortion), which is happening in tetragonal structure.

4.8.2 (1-x)(Na$_{1-y}$K$_y$)(Nb$_{1-z}$Sb$_z$)O$_3$-xBaTiO$_3$, (x=0.1,0.2; y=0.03,0.05; z=0.05,0.1)

The two dimensional charge density distributions show the contour lines around the Na-O bond (figures 3.44 (a)-(d) atoms. No overlapping of charge distribution is observed between Na-O bonding in (001) plane, which indicates that an ionic bonding exists between Na and O atoms. On the other hand, the (002) plane representing the charge density distributions between Nb and O atoms shows electron overlapping between Nb and O atoms (figures 3.45 (a)-(d)). This means that covalent bonding exists between Nb and O atoms. Quantitative analysis of nature of bonding was carried out through one dimensional charge density line profiles analysis along the Na-O and Nb-O bonds. Figures 3.46 (a) and (b) show the one dimensional charge density profiles along Na-O bonding direction and along the Nb-O bonding direction, respectively.

4.8.2.1 Correlation between charge density and piezoelectric properties of (1-x)NKNS-xBT ceramics

The ferroelectric and piezoelectric properties of the (1-x)NKNS-xBT ceramics were analyzed through charge density distribution studies. The ferroelectric properties of (1-x)NKNS-xBT system are attributed to the structural distortion and electronic configuration in the host lattice which allow polarization. For x=0.1 composition, the mid bond electron density values between Nb and O atoms are larger than those of Na and O atoms (Table 3.32). This leads to the covalent nature between Nb and O atoms. The covalent nature is responsible for piezoelectric properties of the prepared samples (Cohen, 1992). With increasing $BaTiO_3$ (x=0.2) content, the crystal structure transforms from tetragonal to cubic phase as evident from the XRD results. Hence, the compositions (x=0.2) show the reduced piezoelectric and ferroelectric properties in the (1-x)NKNS-xBT system.

4.8.3 (1-x)(Na$_{0.5}$Bi$_{0.5}$)TiO$_3$-xBaTiO$_3$, (x=0.00, 0.04, 0.08, 0.12)

4.8.3.1 Correlation between charge density and piezoelectric properties of (1-x)NKNS-xBT ceramics

In titanate based perovskites such as $PbTiO_3$ and $BaTiO_3$, the covalent nature of Ti-O bond is primarily responsible for ferroelectric polarization (Kuroiwa *et al.*, 2001). The ferroelectric properties of undoped $Na_{0.5}Bi_{0.5}TiO_3$ originates due to the lone pair effect of Bi^{3+} at the A-site (Tripathy *et al.*, 2014). The enhanced ferroelectric properties of undoped $Na_{0.5}Bi_{0.5}TiO_3$ ceramic are attributed to the covalent nature between A and O ions. The mid bond electron density value is higher for undoped $Na_{0.5}Bi_{0.5}TiO_3$ (0.5590 e/$Å^3$) than doped samples. Covalency of A-O bonds induces polarization in the structure which lead to ferroelectric property at x=0.00 and x=0.04. At x=0.08, the rhombohedral crystal structure changes to tetragonal due to the more lattice distortion between Ti and O atoms. The bond length between Ti-O bond is higher in x=0.08 (Table 3.33), which helps to increase the distance between off-centered octahedral Ti ion and neighbouring oxygen ions. This leads to enhanced ferroelectric property in x=0.08 composition. Further addition of $BaTiO_3$ in $Na_{0.5}Bi_{0.5}TiO_3$ shows lower ferroelectric property due to low distortion between Ti and O atoms.

4.8.4   (1-x)($K_{0.5}Bi_{0.5}$)TiO_3-x$BaTiO_3$, (x=0.00, 0.08, 0.12)

Figures 3.52 (b)-(c) demonstrate the two dimensional charge density contour maps corresponding to the (001) plane, which show the K/Bi-O bonds. It can be observed that the charge accumulation between K/Bi and O ions in the undoped $K_{0.5}Bi_{0.5}TiO_3$ ceramic reveals covalent nature. This is attributed to the lone pair effect of Bi^{3+} ions. The charge density contours further fade away at the middle portion of the K/Bi-O bond path for increasing Ba^{2+} doping concentrations at x=0.08 and 0.12. This indicates that the bonding between K/Bi and O ions become more ionic for x=0.08 and 0.12. Figures 3.53 (b)-(c) illustrate the two dimensional contour maps on (002) plane, which also show the Ti-O bonds. The two dimensional contour maps on (002) plane show the charge overlap between Ti and O ions along the bonding region, which confirms the covalent nature of bonding between the Ti and O ions. Similar results were reported in previous studies (Sasikumar et al., 2017; Mangaiyarkkarasi and Saravanan, 2017).

The one dimensional electron density profiles of (1-x)KBT-xBT ceramics are plotted along K/Bi-O and Ti-O bonding directions (figures 3.54 (a) and (b)). The bond length variation and charge density values at the mid-bond of K/Bi-O and Ti-O bonds obtained from the MEM analysis are given in Table 3.54. The bond length of Ba-O bond for x=0.00 is 2.7880 Å. The bond length increases from 2.8261 to 2.8265 Å with the addition of Ba^{2+} content for x=0.8 and x=0.12 respectively. Also, the bond length for   Ti-O bond for x=0.00 is 1.9730 Å, which increases with the addition of Ba^{2+} doping in the host lattice. The expansion of bond length with the incorporation of Ba^{2+} is consistent with XRD data. The variations in bond length values with Ba^{2+} doping indicate distortion in the $K_{0.5}Bi_{0.5}TiO_3$ structure and the bond length values are given in Table 3.34.

From Table 3.34, for the concentration x=0.00, the electron density at the bond critical point is found to be 0.4212 e/$Å^3$. The higher value of the mid bond electron density due to the lone pair effect of Bi^{3+} ions is as discussed above and the electron density value decreases with increase of Ba^{2+} concentration. This confirms that the K/Bi-O bond is ionic in nature for x=0.08 and 0.12. Again, from Table 3.34 the mid bond electron density values range from 0.7232 e/$Å^3$ to 0.6265 e/$Å^3$, which fact leads to the covalent behavior between Ti and O ions.

4.8.5 $(1-x)Ba(Zr_{0.2}Ti_{0.8})O_3-x(Ba_{0.7}Ca_{0.3})TiO_3$, (x=0.4, 0.5, 0.6)

It is evident that there is no charge overlap between Ba and O atoms which indicates the ionic nature of bonding between Ba and O (figures 3.56 (b)-(d)). Figures 3.57 (b)-(d) illustrate the two-dimensional contour maps for (002) lattice plane representing the charge density distributions between Ti and O ions. The 2D maps for (002) lattice plane show the charge overlap between Ti and O atoms along the bonding region, which fact confirms the covalent nature of bonding between the Ti and O atoms. Figures 3.58 (a) and (b) represent the one dimensional electron density profiles along   Ba-O and Ti-O bonds respectively. The bond length and mid bond electron density values are given in Table 3.35.

4.8.5.1 Correlation between charge density and piezoelectric properties of (1-x)BZT-xBCT ceramics

In titanate perovskites such as $PbTiO_3$ and $BaTiO_3$, the covalent nature of Ti-O bond is primarily responsible for ferroelectric polarization (Kuroiwa *et al.*, 2001). Table 3.35 shows that the mid-bond electron density values for Ba-O bond vary from 0.1689 $e/Å^3$ to 0.1480 $e/Å^3$, which fact leads to the ionic behavior between Ba and O ions. Again, from Table 3.35, the electron density at the middle of the Ti-O bond is found to be 0.9575 $e/Å^3$ for the composition x=0.5. This confirms that the Ti-O bond is covalent in nature. The piezoelectric constant attains a maximum value of d_{33}~276 pC/N at x=0.5, corresponding to a relatively maximum polarization of P_m ~15.12 $\mu C/cm^2$ and a relatively low coercive field of E_C~3.62 kV/cm. The enhanced piezoelectric and ferroelectric properties can be attributed to the covalent nature of the Ti-O bond.

The charge density distributions between the atoms in the unit cell of all the five series of lead-free ceramics have been analyzed using maximum entropy method (Collins, 1982) (MEM). The nature of bond and the mid bond density values of Na-O, Nb-O, K/Bi-O, Ba-O and Ti-O bonds for the lead-free ceramics are presented in Tables 4.7 and 4.8. The mid bond electron density values reveal the bonding nature between the two atoms. From Tables 4.7 and 4.8, it is seen that the A-O bond is ionic in nature and B-O bond is covalent in nature for all the ceramic solid solutions.

Table 4.7 *Comparison of the mid bond electron density values and nature of bond for Na-O, Nb-O and Ti-O bonds for synthesized lead-free solid solutions.*

Sample	Composition	Na-O bond		Nb-O bond	
		Mid bond electron density $(e/Å^3)$	Relative nature of bond	Mid bond electron density $(e/Å^3)$	Relative nature of bond
I	x=0.1, y=0.01	0.0697	Ionic	1.0558	Covalent
	x=0.1, y=0.05	0.0715	Ionic	0.6317	Covalent
	x=0.2, y=0.01	0.1223	Ionic	1.3644	Covalent
	x=0.2, y=0.05	0.0963	Ionic	1.3406	Covalent
II	x=0.1, y=0.03, z=0.05	0.0572	Ionic	0.9775	Covalent
	x=0.1, y=0.05, z=0.05	0.0604	Ionic	1.0133	Covalent
	x=0.2, y=0.03, z=0.1	0.1092	Ionic	1.4875	Covalent
	x=0.2, y=0.05, z=0.1	0.1198	Ionic	1.4886	Covalent

I - $(1-x)(Na_{1-y}K_y)NbO_3-xBaTiO_3$

II - $(1-x)(Na_{1-y}K_y)(Nb_{1-z}Sb_z)O_3-xBaTiO_3$

Table 4.8 *Comparison of the mid bond electron density values and nature of bond for Ba-O, K/Bi-O and Ti-O bonds for synthesized lead-free solid solutions.*

Sample	Composition	Ba-O bond		K/Bi-O bond		Ti-O bond	
		Mid bond electron density $(e/Å^3)$	Relative nature of bond	Mid bond electron density $(e/Å^3)$	Relative nature of bond	Mid bond electron density $(e/Å^3)$	Relative nature of bond
III	x=0.00	0.5590	Covalent			0.1919	Covalent
	x=0.04	0.3139	Ionic			0.7554	Covalent
	x=0.08	0.2950	Ionic			0.7072	Covalent
	x=0.12	0.2650	Ionic			0.6049	Covalent
IV	x=0.00			0.4212	Ionic	0.7232	Covalent
	x=0.08			0.3614	Ionic	0.8683	Covalent
	x=0.12			0.2967	Ionic	0.6265	Covalent
V	x=0.4	0.1681	Ionic			0.6998	Covalent
	x=0.5	0.1689	Ionic			0.9575	Covalent
	x=0.6	0.1480	Ionic			0.6150	Covalent

III - $(1-x)(Na_{0.5}Bi_{0.5})TiO_3-xBaTiO_3$

IV - $(1-x)(K_{0.5}Bi_{0.5})TiO_3-xBaTiO_3$

V - $(1-x)Ba(Zr_{0.2}Ti_{0.8})O_3-x(Ba_{0.7}Ca_{0.3})TiO_3$

References

[1] Benlahrache M.T., Benhamla N., Achour S., J. Eur. Ceram. 24, 1493 (2004). https://doi.org/10.1016/S0955-2219(03)00577-6

[2] Bhattacharya K., Ravichandran G., Acta Mater. 51, 5941 (2003). https://doi.org/10.1016/j.actamat.2003.08.001

[3] Coondoo I., Panwar N., Amorın H., Alguero M., Kholkin A.L., J. Appl. Phys. 113, 214107 (2013). https://doi.org/10.1063/1.4808338

[4] Cohen R.E., Nature, London 358, 136-138 (1992). https://doi.org/10.1038/358136a0

[5] Chang-lin G., Yu-qin W., Tian-Bao W., Wuli Xuebao 31, 1119 (1982).

[6] Ehmke M.C., Daniels J., Glaum J., Hoffman M., Blendell J.E., Bowman K.J., J. Appl. Phys. 112, 114108 (2012). https://doi.org/10.1063/1.4768273

[7] Fu J., Zuo R.Z., Liu Y., J. Alloy. Compd. 493, 197 (2010). https://doi.org/10.1016/j.jallcom.2009.12.052

[8] Koops C.G., Phys. Rev. B 83, 108 (1951).

[9] Kuroiwa Y., Aoyagi S., Sawada A., Harada J., Nishibori E., Takata M., Sakata M., Phys. Rev. Lett. 87, 217601 (2001). https://doi.org/10.1103/PhysRevLett.87.217601

[10] Mangaiyarkkarasi J., Saravanan R., J Mater. Sci: Mater. Electron 28(3), 2624-2633 (2017). https://doi.org/10.1007/s10854-016-5839-9

[11] Mishra P., Sonia Kumar P., Ceram. Int. 40, 14149 (2014). https://doi.org/10.1016/j.ceramint.2014.06.001

[12] Momma K., Izumi F., VESTA 3 for three-dimensional visualization of crystal, volumetric and morphology data, J. Appl. Crystallogr. 44, 1272-1276 (2011). https://doi.org/10.1107/S0021889811038970

[13] Patil D.R., Lokare S.A., Devan R.S., Chougule S.S., Kanamadi C.M, Kolekar Y.D., Chougule B.K., Mater. Chem. Phys 104, 254 (2007). https://doi.org/10.1016/j.matchemphys.2007.02.027

[14] Petříček V., Dusek M., Palatinus J., The crystallographic computing system (Institute of Physics) (Czech Republic, Praha, 2006).

[15] Qu Y., Shan D., Song J., Mater. Sci. Eng. B121, 148 (2005). https://doi.org/10.1016/j.mseb.2005.03.023

[16]   Rietveld H.M., J. Appl. Crystallogr. 2, 65 (1969).
https://doi.org/10.1107/S0021889869006558

[17]   Sasikumar S., Saravanan R., Saravanakumar S., Aravinth K., J Mater. Sci Mater.
Electron. 28, 9950-9963 (2017). https://doi.org/10.1007/s10854-017-6753-5

[18]   Shah J., Kotnala R.K., J. Mater. Chem. A1, 8601-8608 (2013).
https://doi.org/10.1039/c3ta11845b

[19]   Shannon R.D., Acta Cryst. A32, 751-767 (1976).
https://doi.org/10.1107/S0567739476001551

[20]   Shi S., Zuo R., Lu S., Xu Z., Wang X., Li L., Curr. Appl. Phys.11, s120 (2011).
https://doi.org/10.1016/j.cap.2011.03.019

[21]   Sridevi S., Subrat Kumar K., Pawan K., Ceram. Int. 41(9), 10710 (2015).
https://doi.org/10.1016/j.ceramint.2015.05.005

[22]   Tripathy S., Mishra K., Sen S., Pradhan D., J. Am. Ceram. Soc. 97(6), 1846
(2014). https://doi.org/10.1111/jace.12838

[23]   Wyckoff R.W.G., C. Structures, John Wiley, New York, (1963).

[24]   Xu C., Lin D., Kwok K.W., Solid. State. Sci. 10, 934 (2008).
https://doi.org/10.1016/j.solidstatesciences.2007.11.003

[25]   Yang C.-F., Chen K.-N., Wu C.-C., Hsun Lin Y., Ceram. Int. 42(7), 8932 (2016).
https://doi.org/10.1016/j.ceramint.2016.02.150

[26]   Yingze Z., L. Qianru L., L. Wenfeng L., W. Danyang W., RSC Adv. 6, 14084
(2016).

[27]   Zeng J., Zheng L., Li G., Cao Z., Zhao K., Yin Q., Politova E.D., J. Alloy. Compd.
509, 5858-5862 (2011). https://doi.org/10.1016/j.jallcom.2011.02.152

Chapter 5

Conclusion

Abstract

This chapter gives the conclusion of the findings of the reported work.

Keywords: Conclusions, Structural Analysis, Elemental Analysis, Dielectric Constant, Ferro Electric Measurement, Piezo Electric Measurement, Electron Density Analysis

5. Conclusion

The following conclusions are made from the present studies on the lead-free ceramics.

i) Structural analysis

- The phase formation and structural analysis of the prepared lead-free solid solutions have been carried out by X-ray diffraction and powder XRD profile refinement study.
- $(1-x)$NKN-xBT and $(1-x)$NKNS-xBT solid solutions show the presence of a morphotropic phase boundary (MPB) between tetragonal and cubic phase in the composition range x=0.1 to 0.2 respectively.
- The crystal structure of the $(1-x)$NBT-xBT ceramics changes with increasing $BaTiO_3$ content and a morphotropic phase transition from rhombohedral to tetragonal phases has been identified at x=0.08.
- The XRD profile refinement of $(1-x)$BZT-xBCT ceramics suggested the coexistence of mixed crystallographic phases of tetragonal and rhombohedral at the composition x=0.5.

ii) Microstructure, elemental and optical analysis

- SEM micrographs of the prepared lead-free ceramics show that the size and shape of the particles apparently changed with the addition of $BaTiO_3$ content.
- The average particle size gradually decreases with the addition of $BaTiO_3$ content.
- No additional impurities were detected in the EDS spectrum.
- The optical band gap energy of solid solution ceramics decreases with an increase of the $BaTiO_3$ content.

iii) Dielectric properties

- In (1-x)NKN-xBT and (1-x)NKNS-xBT solid solutions, the temperature dependent dielectric studies show a tetragonal to distorted cubic phase transition.
- In (1-x)KBT-xBT ceramics, frequency dependence of dielectric properties has been investigated at different frequencies.
- In (1-x)BZT-xBCT ceramics, the Curie temperature (T_C) shifted towards higher temperature side with the increase of x value.

iv) Ferroelectric and piezoelectric properties

- In (1-x)NKN-xBT and (1-x)NKNS-xBT solid solutions, the remnant polarization decreases with addition of $BaTiO_3$ content due to the phase transition from tetragonal to cubic.
- Enhanced ferroelectric property is observed in undoped $Na_{0.5}Bi_{0.5}TiO_3$ and $K_{0.5}Bi_{0.5}TiO_3$ ceramics due to the lone pair effect of electron in the Bi^{3+} ions.
- In (1-x)NBT-xBT solid solution, the piezoelectric property is found to increase at x=0.08 composition due to near the MPB region.
- In (1-x)BZT-xBCT solid solution, the composition x=0.5 exhibits maximum polarization at room temperature, implying an MPB like behaviour.
- The incorporation of a small amount of $BaTiO_3$ (x=0.1) improves the piezoelectric properties of (1-x)NKN-xBT and (1-x)NKNS-xBT solid solutions.
- In (1-x)BZT-xBCT solid solution, the d_{33} value is higher at x=0.5 (d_{33}=276 pC/N), which is attributed to the coexistence of rhombohedral and tetragonal phases.

v) Charge density analysis

- The charge density distribution analysis reveals the bonding nature between the atoms of the prepared lead-free solid solutions.
- The A-O bond is ionic in nature and B-O bond is covalent in nature for all the ceramic solid solutions.
- The presence of more charges in the mid bond region of Nb-O atoms is attributed to the enhanced ferroelectric and piezoelectric properties.

The prepared solid solutions of lead-free ceramics have been analyzed through various experimental techniques and charge density distributions. The ferroelectric and piezoelectric properties have been correlated to the charge density distributions in the unit cell. It is found that most improved dielectric, ferroelectric and piezoelectric properties occur near the MPB region. Among all the compositions of (1-x)BZT-xBCT, the

Materials Research Forum LLC
doi: http://dx.doi.org/10.21741/9781945291951

composition x=0.5 shows the best piezoelectric and ferroelectric properties which makes the material suitable for industrial applications and which might be a promising lead-free piezoelectric material for various kinds of actuators and transducers.

Keyword Index

About the Author

Dr Ramachandran Saravanan, has been associated with the Department of Physics, The Madura College, affiliated with the Madurai Kamaraj University, Madurai, Tamil Nadu, India from the year 2000. He is the head of the Research Centre and PG department of Physics. He worked as a research associate during 1998 at the Institute of Materials Research, Tohoku University, Sendai, Japan and then as a visiting researcher at Centre for Interdisciplinary Research, Tohoku University, Sendai, Japan up to 2000.

Earlier, he was awarded the Senior Research Fellowship by CSIR, New Delhi, India, during Mar. 1991 - Feb.1993; awarded Research Associateship by CSIR, New Delhi, during 1994 – 1997. Then, he was awarded a Research Associateship again by CSIR, New Delhi, during 1997- 1998. Later he was awarded the Matsumae International Foundation Fellowship in1998 (Japan) for doing research at a Japanese Research Institute (not availed by him due to the simultaneous occurrence of other Japanese employment).

He has guided twelve Ph.D. scholars as of 2018, and about six researchers are working under his guidance on various research topics in materials science, crystallography and condensed matter physics. He has published around 150 research articles in reputed Journals, mostly International, apart from around 50 presentations in conferences, seminars and symposia. He has also guided around 60 M.Phil. scholars and an equal number of PG students for their projects. He has attracted government funding in India, in the form of Research Projects. He has completed two CSIR (Council of Scientific and Industrial Research, Govt. of India), one UGC (University Grants Commission, India) and one DRDO (Defense Research and Development Organization, India) research projects successfully and is proposing various projects to Government funding agencies like CSIR, UGC and DST.

He has written 12 books in the form of research monographs including; "Experimental Charge Density - Semiconductors, oxides and fluorides" (ISBN-13: 978-3-8383-8816-8; ISBN-10:3-8383-8816-X), "Experimental Charge Density - Dilute Magnetic Semiconducting (DMS) materials" (ISBN-13: 978-3-8383-9666-8; ISBN-10: 3-8383-9666-9) and "Metal and Alloy Bonding - An Experimental Analysis" (ISBN -13: 978-1-4471-2203-6). He has committed to write several books in the near future.

His expertise includes various experimental activities in crystal growth, materials science, crystallographic, condensed matter physics techniques and tools as in slow evaporation, gel, high temperature melt growth, Bridgman methods, CZ Growth, high vacuum sealing etc. He and his group are familiar with various equipment such as: different types of cameras; Laue, oscillation, powder, precession cameras; Manual 4-circle X-ray diffractometer, Rigaku 4-circle automatic single crystal diffractometer, AFC-5R and AFC-7R automatic single crystal diffractometers, CAD-4 automatic single crystal diffractometer, crystal pulling instruments, and other crystallographic, material science related instruments. He and his group have sound computational capabilities on different types of computers such as: IBM – PC, Cyber180/830A – Mainframe, SX-4 Supercomputing system – Mainframe. He is familiar with various kind of software related to crystallography and materials science. He has written many computer software programs himself as well. Around twenty of his programs (both DOS and GUI versions) have been included in the SINCRIS software database of the International Union of Crystallography.

www.ingramcontent.com/pod-product-compliance
Lightning Source LLC
Chambersburg PA
CBHW071233210326
41597CB00016B/2034